Cell Biology
Fundamentals & Applications

Cell Biology
Fundamentals & Applications

Arunima Mukherjee

Oxford Book Company
Jaipur ▪ New Delhi

Published By

Oxford Book Company

Regd. Off. "Bony Residency", Gate No. 2
Opp. Tilak Public School, Vishweswariya Nagar,
Gopalpura Road, Jaipur - 302018, India
Ph: 0141 - 2761280, Telefax: 2761381
email - oxfordbookjpr@gmail.com
website - oxfordbookcompany.com

Delhi Office: 102, 1st Floor, Satyam House,
4327/3, Ansari Road, Darya Ganj,
New Delhi - 110002, India
Ph: 011-45652440

Distributed by
KEYA INTERNATIONAL PRESS LLP
E-mail: keyainternationalpressllp@gmail.com

Published in India, 2024

ISBN : 9788189473501

Typeset by Shivangi Computers, Jaipur

Printed and bound in India by Replika Press Pvt. Ltd.

Contents

Preface

Cell biology is an academic discipline that studies cells – their physiological properties, their structure, the organelles they contain, interactions with their environment, their life cycle, division and death. This is done both on a microscopic and molecular level. Cell biology research extends to both the great diversity of single-celled organisms like bacteria and the many specialised cells in multicellular organisms like humans.

Knowing the composition of cells and how cells work is fundamental to all of the biological sciences. Appreciating the similarities and also differences between cell types is particularly important to the fields of cell and molecular biology. These fundamental similarities and differences provide a unifying theme, allowing the principles learned from studying one cell type to be extrapolated and generalised to other cell types.

This book sets forth the current understanding of cell biology and explores the intriguing implications and possibilities of the great deal that remains unknown. By extracting the fundamental concepts from this ever-growing field, the book tell the story of cell biology, and create a coherent framework through which non-expert readers may approach the subject. Written in clear and concise language, the book is enjoyable to read, and it provides a clear sense of the excitement of modern biology.

— *Arunima Mukherjee*

1

Structure and Functions of Cell

The cell is the structural and functional unit of all known living organisms. It is the smallest unit of an organism that is classified as living, and is sometimes called the building block of life. Some organisms, such as most bacteria, are unicellular (consist of a single cell). Other organisms, such as humans, are multicellular. (Humans have an estimated 100 trillion or 1014 cells; a typical cell size is 10 µm; a typical cell mass is 1 nanogram.)

The largest known cell is an ostrich egg. In 1837 before the final cell theory was developed, a Czech Jan Evangelista Purkyne observed small "granules" while looking at the plant tissue through a microscope. The cell theory, first developed in 1839 by Matthias Jakob Schleiden and Theodor Schwann, states that all organisms are composed of one or more cells.

All cells come from preexisting cells. Vital functions of an organism occur within cells, and all cells contain the hereditary information necessary for regulating cell functions and for transmitting information to the next generation of cells.

PLANT CELL

Plants are unique among the eukaryotes, organisms whose cells have membrane-enclosed nuclei and organelles, because they can manufacture their own food. Chlorophyll, which gives plants their green color, enables them to use sunlight to convert water and carbon dioxide into sugars and carbohydrates, chemicals the cell uses for fuel.

Like the fungi, another kingdom of eukaryotes, plant cells have retained the protective cell wall structure of their prokaryotic ancestors. The basic plant cell shares a similar construction motif with the typical eukaryote cell, but does not have centrioles, lysosomes, intermediate filaments, cilia, or flagella, as does the animal cell. Plant cells do, however, have a number of other specialized structures, including a rigid cell wall, central vacuole, plasmodesmata, and chloroplasts. Although plants (and their typical cells) are non-motile, some species produce gametes that do exhibit flagella and are, therefore, able to move about.

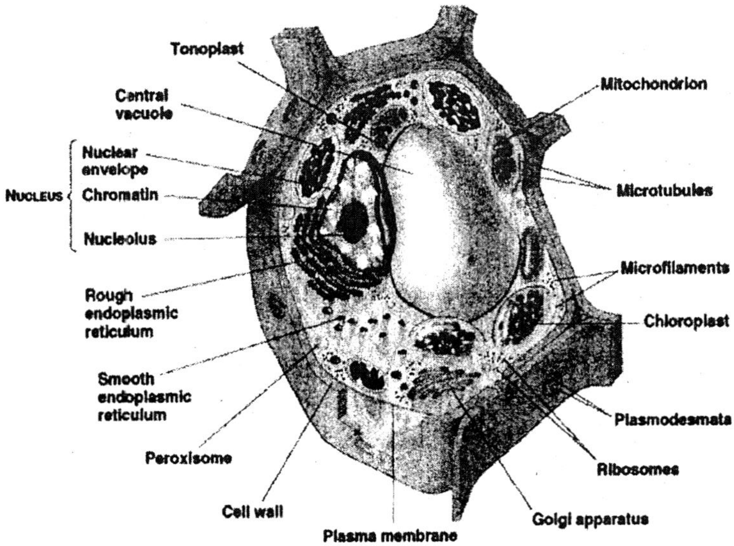

Anatomy of the plant cell

Plants can be broadly categorized into two basic types: vascular and nonvascular. Vascular plants are considered to be more advanced than nonvascular plants because they have evolved specialized tissues, namely xylem, which is involved in structural support and water conduction, and phloem, which functions in food conduction. Consequently, they also possess roots, stems, and leaves, representing a higher form of organization that is characteristically absent in plants lacking vascular tissues.

The nonvascular plants, members of the division Bryophyta, are usually no more than an inch or two in height because they do not have adequate support, which is provided by vascular tissues to other plants, to grow bigger. They also are more dependent on the environment that surrounds them to maintain appropriate amounts of moisture and, therefore, tend to inhabit damp, shady areas.

It is estimated that there are at least 260,000 species of plants in the world today. They range in size and complexity from small, nonvascular mosses to giant sequoia trees, the largest living organisms, growing as tall as 330 feet (100 meters). Only a tiny percentage of those species are directly used by people for food, shelter, fiber, and medicine.

Nonetheless, plants are the basis for the Earth's ecosystem and food web, and without them complex animal life forms (such as humans) could never have evolved. Indeed, all living organisms are dependent either directly or indirectly on the energy produced by photosynthesis, and the byproduct of this process, oxygen, is essential to animals. Plants also reduce the amount of carbon dioxide present in the atmosphere, hinder soil erosion, and influence water levels and quality.

Plants exhibit life cycles that involve alternating generations of diploid forms, which contain paired chromosome sets in their cell nuclei, and haploid forms, which only possess a single set. Generally these two forms of a plant are very dissimilar in appearance. In higher plants, the diploid generation, the members of which are known as sporophytes due to their ability to produce spores, is usually dominant and more recognizable than the haploid gametophyte generation. In Bryophytes, however, the gametophyte form is dominant and physiologically necessary to the sporophyte form.

Animals are required to consume protein in order to obtain nitrogen, but plants are able to utilize inorganic forms of the element and, therefore, do not need an outside source of protein. Plants do, however, usually require significant amounts of water, which is needed for the photosynthetic process, to maintain cell structure and facilitate growth, and as a means of bringing nutrients to plant cells. The amount of nutrients needed by plant species varies significantly, but nine elements are generally considered to be necessary in relatively large amounts. Termed macroelements, these nutrients include calcium, carbon, hydrogen, magnesium, nitrogen, oxygen, phosphorus, potassium, and sulfur. Seven microelements, which are required by plants in smaller quantities, have also been identified: boron, chlorine, copper, iron, manganese, molybdenum, and zinc.

Thought to have evolved from the green algae, plants have been around since the early Paleozoic era, more than 500 million years ago. The earliest fossil evidence of land plants dates to the Ordovician Period (505 to 438 million years ago). By the Carboniferous Period, about 355 million years ago, most of the Earth was covered by forests of primitive vascular plants, such as lycopods (scale trees) and gymnosperms (pine trees, ginkgos). Angiosperms, the flowering plants, didn't develop until the end of the Cretaceous Period, about 65 million years ago-just as the dinosaurs became extinct.

The plant body is divided into several organs: roots, stems, and leaves. The leaves are the primary photosynthetic organs of plants, serving as key sites where energy from light is converted into chemical energy. Similar to the other organs of a plant, a leaf is comprised of three basic tissue systems, including the dermal, vascular, and ground tissue systems. These three motifs are continuous throughout an entire plant, but their properties vary significantly based upon the organ type in which they are located.

ANIMAL CELL

Animal cells are typical of the eukaryotic cell, enclosed by a plasma membrane and containing a membrane-bound nucleus and organelles. Unlike the eukaryotic cells of plants and fungi, animal cells do not have a cell wall. This feature was lost in the distant past by the single-celled organisms that gave rise to the kingdom Animalia. Most cells, both animal and plant, range in size between 1 and 100 micrometers and are thus visible only with the aid of a microscope.

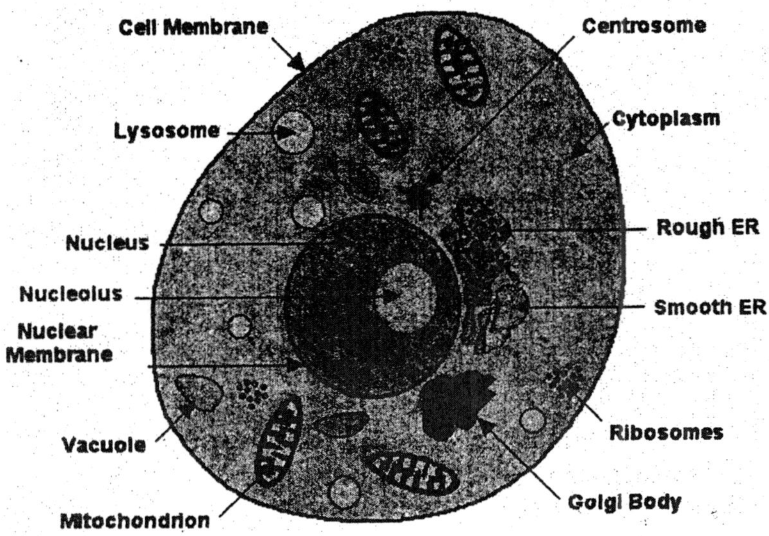

Anatomy of the animal cell

The lack of a rigid cell wall allowed animals to develop a greater diversity of cell types, tissues, and organs. Specialized cells that formed nerves and muscles-tissues impossible for plants to evolve-gave these organisms mobility. The ability to move about by the use of specialized muscle tissues is a hallmark of the animal world, though a few animals, primarily sponges, do not possess differentiated tissues. Notably, protozoans locomote, but it is only via nonmuscular means, in effect, using cilia, flagella, and pseudopodia.

The animal kingdom is unique among eukaryotic organisms because most animal tissues are bound together in an extracellular matrix by a triple helix of protein known as collagen. Plant and fungal cells are bound together in tissues or aggregations by other molecules, such as pectin. The fact that no other organisms utilize collagen in this manner is one of the indications that all animals arose from a common unicellular ancestor. Bones, shells, spicules, and other hardened structures are formed when the collagen-containing extracellular matrix between animal cells becomes calcified.

Animals are a large and incredibly diverse group of organisms. Making up about three-quarters of the species on Earth, they run the gamut from corals and jellyfish to ants, whales, elephants, and, of course, humans. Being mobile has given animals, which are capable of sensing and responding to their environment, the flexibility to adopt many different modes of feeding, defense, and reproduction. Unlike plants, however, animals are unable to manufacture their own food, and therefore, are always directly or indirectly dependent on plant life.

Most animal cells are diploid, meaning that their chromosomes exist in homologous pairs. Different chromosomal ploidies are also, however, known to occasionally occur. The proliferation of animal cells occurs in a variety of ways. In instances of sexual reproduction, the cellular process of meiosis is first necessary so that haploid daughter cells, or gametes, can be produced. Two haploid cells then fuse to form a diploid zygote, which develops into a new organism as its cells divide and multiply.

The earliest fossil evidence of animals dates from the Vendian Period (650 to 544 million years ago), with coelenterate-type creatures that left traces of their soft bodies in shallow-water sediments. The first mass extinction ended that period, but during the Cambrian Period which followed, an explosion of new forms began the evolutionary radiation that produced most of the major groups, or phyla, known today. Vertebrates (animals with backbones) are not known to have occurred until the early Ordovician Period (505 to 438 million years ago).

Cells were discovered in 1665 by British scientist Robert Hooke who first observed them in his crude (by today's standards) seventeenth century optical microscope. In fact, Hooke coined the term "cell", in a biological context, when he described the microscopic structure of cork like a tiny, bare room or monk's cell. The microscope has been a fundamental tool in the field of cell biology and is often used to observe living cells in culture

COMPONENTS OF CELL

Cell Wall

One of the most important distinguishing features of plant cells is the presence of a cell wall. The relative rigidity of the cell wall renders plants sedentary, unlike animals, whose lack of this type of structure allows their cells more flexibility, which is necessary for locomotion. The plant cell wall serves a variety of functions. Along with protecting the intracellular contents, the structure bestows rigidity to the plant, provides a porous medium for the circulation and distribution of water, minerals, and other nutrients, and houses specialized molecules that regulate growth and protect the plant from disease.

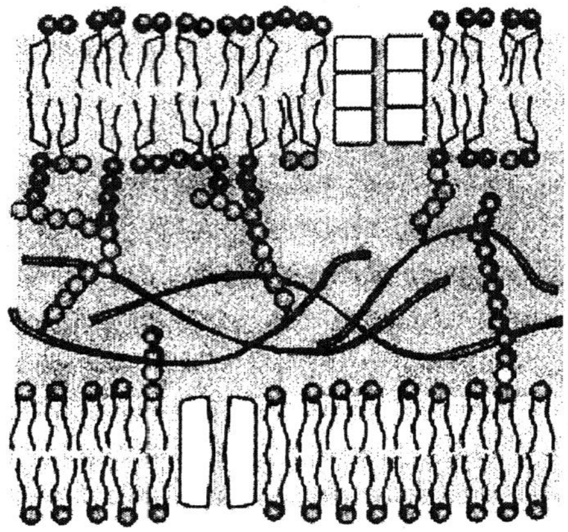

Cell wall structure of Corynebacterianeae

Cell walls are significantly thicker than plasma membranes and were visible even to early microscopists, including Robert Hooke, who originally identified the structures in a sample of cork, and then coined the term cells in the 1660s. The thickness, as well as the composition and organization, of cell walls can vary significantly. Many plant cells have both a primary cell wall, which accommodates the cell as it grows, and a secondary cell wall they develop inside the primary wall after the cell has stopped growing. The primary cell wall is thinner and more pliant than the secondary cell wall, and is sometimes retained in an unchanged or slightly modified state without the addition of the secondary wall, even after the growth process has ended.

The main chemical components of the primary plant cell wall include cellulose, a complex carbohydrate made up of several thousand glucose molecules linked end to end. In addition, the cell wall contains two groups of branched polysaccharides, the pectins and cross-linking glycans. Organized into a network with the cellulose microfibrils, the cross-linking glycans increase the tensile strength of the cellulose, whereas the coextensive network of pectins provides the cell wall with the ability to resist compression. In addition to these networks, a small amount of protein can be found in all plant primary cell walls. Some of this protein is thought to increase mechanical strength and part of it consists of enzymes, which initiate reactions that form, remodel, or breakdown the structural networks of the wall. Such changes in the cell wall directed by enzymes are particularly important for fruit to ripen and leaves to fall in autumn.

The secondary plant cell wall, which is often deposited inside the primary cell wall as a cell matures, sometimes has a composition nearly identical to that of the earlier-developed wall. More commonly, however, additional substances, especially lignin, are found in the secondary wall. Lignin is the general name for a group of polymers of aromatic alcohols that are hard and impart considerable strength to the structure of the secondary wall. Lignin is what provides the favorable characteristics of wood to the fiber cells of woody tissues and is also common in the secondary walls of xylem vessels, which are central in providing structural support to plants. Lignin also makes plant cell walls less vulnerable to attack by fungi or bacteria, as do cutin, suberin, and other waxy materials that are sometimes found in plant cell walls.

A specialized region associated with the cell walls of plants, and sometimes considered an additional component of them, is the middle lamella. Rich in pectins, the middle lamella is shared by neighboring cells and cements them firmly together. Positioned in such a manner, cells are able to communicate with one another and share their contents through special conduits. Termed plasmodesmata, these small passages penetrate the middle lamella as well as the primary and secondary cell walls, providing pathways for transporting cytoplasmic molecules from one cell to another.

Centrioles

Found only in animal cells, these paired organelles are typically located together near the nucleus in the centrosome, a granular mass that serves as an organizing center for microtubules.

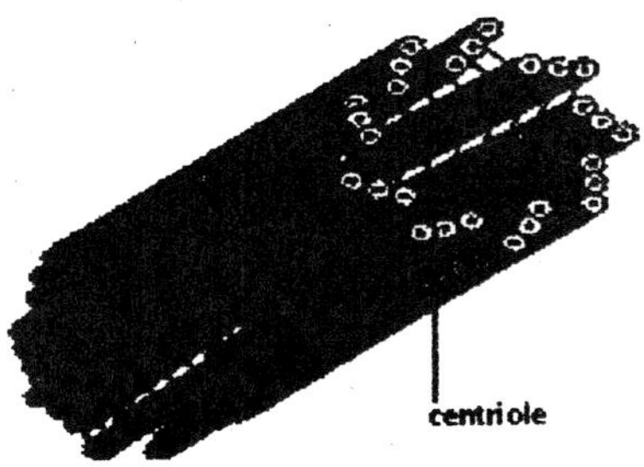

Centriole structure

Centrioles play a notable role in cell division. During interphase of an animal cell, the centrioles and other components of the centrosome are duplicated, though scientists are not yet sure how this duplication takes place. At first the two pairs of centrioles remain in close proximity to each other, but as mitosis initiates, the original centrosome divides and the pairs are split up so that one set of centrioles is located in each of the new microtubule-organizing centers. These new centers radiate microtubules in star-shaped clusters known as asters. As the asters move to opposing poles of the cells, the microtubules, with the help of the centrioles, become organized into a spindle-shaped formation that spans the cell. These spindle fibers act as guides for the alignment of the chromosomes as they separate later during the process of cell division.

Though centrioles play a role in the mitosis of animal cells, plant cells are able to reproduce without them. Researchers have, therefore, been very interested in determining exactly how important the organelles really are. Studies have shown that certain animal cells, particularly female gametes (oocytes), can successfully divide even when their centrioles are destroyed. Some investigators have also found, however, that the absence of centrioles in animal cells is associated with an increased number of divisional errors and substantial delays in the mitotic process, especially before chromosome segregation. Consequently, it has been suggested that centrioles evolved as a refinement of the cell, making mitosis a much more efficient and less error-prone process.

In cells that feature cilia or flagella, basal bodies, which exhibit the same structural form as centrioles, are present. These assemblies are located, however, near the cell surface at the base of each cilium or flagellum, rather than in the centrosome near the nucleus. Basal bodies are anchored in their cytoplasmic locations by what is called a rootlet system in the cell. In some organisms, such as the unicellular Chlamydomonas, basal bodies change their location and are functionally converted to centrioles before the mitotic process.

Chloroplasts

One of the most widely recognized and important characteristics of plants is their ability to conduct photosynthesis, in effect, to make their own food by converting light energy into chemical energy. This process occurs in almost all plant species and is carried out in specialized organelles known as chloroplasts. All of the green structures in plants, including stems and unripened fruit, contain chloroplasts, but the majority of photosynthesis activity in most plants occurs in the leaves.

On the average, the chloroplast density on the surface of a leaf is about one-half million per square millimeter.

Plant cell chloroplast structure

Chloroplasts are one of several different types of plastids, plant cell organelles that are involved in energy storage and the synthesis of metabolic materials. The colorless leucoplasts, for instance, are involved in the synthesis of starch, oils, and proteins. Yellow-to-red colored chromoplasts manufacture carotenoids, and the green colored chloroplasts contain the pigments chlorophyll a and chlorophyll b, which are able to absorb the light energy needed for photosynthesis to occur. All plastids develop from tiny organelles found in the immature cells of plant meristems (undifferentiated plant tissue) termed proplastids, and those of a particular plant species all contain copies of the same circular genome. The disparities between the various types of plastids are based upon the needs of the differentiated cells they are contained in, which may be influenced by environmental conditions, such as whether light or darkness surrounds a leaf as it grows.

The ellipsoid-shaped chloroplast is enclosed in a double membrane and the area between the two layers that make up the membrane is called the intermembrane space. The outer layer of the double membrane is much more permeable than the inner layer, which features a number of embedded membrane transport proteins. Enclosed by the chloroplast membrane is the stroma, a semi-fluid material that contains dissolved enzymes and comprises most of the chloroplast's volume. Since, like mitochondria, chloroplasts possess their own genomes (DNA), the stroma contains chloroplast DNA and special ribosomes and RNAs as well. In higher plants, lamellae, internal membranes with stacks (each termed a granum) of closed hollow disks called thylakoids, are also usually dispersed throughout the stroma. The numerous thylakoids in each stack are thought to be connected via their lumens (internal spaces).

Light travels as packets of energy called photons and is absorbed in this form by light-absorbing chlorophyll molecules embedded in the thylakoid disks. When these chlorophyll molecules absorb the photons, they emit electrons, which they obtain from water (a process that results in the release of oxygen as a byproduct). The movement of the electrons causes hydrogen ions to be propelled across the membrane surrounding the thylakoid stack, which consequently initiates the formation of an electrochemical gradient that drives the stroma's production of adenosine triphosphate (ATP). ATP is the chemical energy "currency" of the cell that powers the cell's metabolic activities. In the stroma, the light-independent reactions of photosynthesis, which involve carbon fixation, occur, and low-energy carbon dioxide is transformed into a high-energy compound like glucose.

Plant cells are remarkable in that they have two organelles specialized for energy production: chloroplasts, which create energy via photosynthesis, and mitochondria, which generate energy through respiration, a particularly important process when light is unavailable. Like the mitochondrion, the chloroplast is different from most other organelles because it has its own DNA and reproduces independently of the cell in which it is found; an apparent case of endosymbiosis. Scientists hypothesize that millions of years ago small, free-living prokaryotes were engulfed, but not consumed, by larger prokaryotes, perhaps because they were able to resist the digestive enzymes of the engulfing organism. According to DNA evidence, the eukaryotic organisms that later became plants likely added the photosynthetic pathway in this way, by acquiring a photosynthetic bacterium as an endosymbiont.

As suggested by this hypothesis, the two organisms developed a symbiotic relationship over time, the larger organism providing the smaller with ample nutrients and the smaller organism providing ATP molecules to the larger one. Eventually, the larger organism developed into the eukaryotic cell, the smaller organism into the chloroplast. Nonetheless, there are a number of prokaryotic traits that chloroplasts continue to exhibit. Their DNA is circular, as it is in the prokaryotes, and their ribosomes and reproductive methods (binary fission) are more like those of the prokaryotes.

Cilia and Flagella

Cilia and flagella are motile cellular appendages found in most microorganisms and animals, but not in higher plants.. In multicellular organisms, cilia function to move a cell or group of cells or to help transport fluid or materials past them. The respiratory tract in humans is lined with cilia that keep inhaled dust, smog, and potentially harmful microorganisms from entering the lungs. Among other tasks, cilia also generate water currents to carry food and oxygen past the gills of clams and transport food through the digestive systems of snails. Flagella are found primarily on gametes, but create the water currents necessary for respiration and circulation in sponges and coelenterates as well. For single-celled eukaryotes, cilia and flagella are essential for the locomotion of individual organisms. Protozoans belonging to the phylum Ciliophora are covered with cilia, while flagella are a characteristic of the protozoan group Mastigophora.

In eukaryotic cells, cilia and flagella contain the motor protein dynein and microtubules, which are composed of linear polymers of globular proteins called tubulin. The core of each of the structures is termed the axoneme and contains two central microtubules that are surrounded by an outer ring of nine doublet microtubules. One full microtubule and one partial microtubule, the latter of which shares a tubule wall with the other microtubule, comprise each doublet microtubule. Dynein molecules are located around the circumference of the axoneme at regular intervals along its length where they bridge the gaps between adjacent microtubule doublets.

A plasma membrane surrounds the entire axoneme complex, which is attached to the cell at a structure termed the basal body (also known as a kinetosome). Basal bodies maintain the basic outer ring structure of the axoneme, but each of the nine sets of circumferential filaments is composed of three microtubules, rather than a doublet of microtubules. Thus, the basal body is structurally identical to the centrioles that are found in the centrosome located near the nucleus of the cell. In some organisms, such as the unicellular Chlamydomonas, basal bodies are locationally and functionally altered into centrioles and their flagella resorbed before cell division.

Eukaryotic cilia and flagella are generally differentiated based on size and number: cilia are usually shorter and occur together in much greater numbers than flagella, which are often solitary. The structures also exhibit somewhat different types of motion, though in both cases movement is generated by the activation of dynein and the resultant bending of the axoneme. The movement of cilia is often described as whip-like, or compared to the breast stroke in swimming. Adjacent cilia move almost simultaneously (but not quite), so that in groups of cilia, wave-like patterns of motion occur. Flagella, however, exhibit a smooth, independent undulatory type of movement in eukaryotes. Prokaryotic flagella, which have a completely different structure built from the protein flagellin, move in a rotating fashion powered by the basal motor.

Defects in the cilia and flagella of human cells are associated with some notable medical problems. For example, a hereditary condition known as Kartagener's syndrome is caused by problems with the dynein arms that extend between the microtubules present in the axoneme, and is characterized by recurrent respiratory infections related to the inability of cilia in the respiratory tract to clear away bacteria or other materials. The disease also results in male sterility due to the inability of sperm cells to propel themselves via flagella. Damage to respiratory cilia may also be acquired rather than inherited and is most commonly linked to smoking cigarettes. Bronchitis, for instance, is often triggered by a build-up of mucus and tar in the lungs that cannot be properly removed due to smoking-related impairment of cilia.

Endoplasmic Reticulum

The endoplasmic reticulum (ER) is a network of flattened sacs and branching tubules that extends throughout the cytoplasm in plant and animal cells. These sacs and tubules are all interconnected by a single continuous membrane so that the organelle has only one large, highly convoluted and complexly arranged lumen (internal space). Usually referred to as the endoplasmic reticulum cisternal space, the lumen of the organelle often takes up more than 10 percent of the total volume of a cell. The endoplasmic reticulum membrane allows molecules to be selectively transferred between the lumen and the cytoplasm, and since it is connected to the double-layered nuclear envelope, it further provides a pipeline between the nucleus and the cytoplasm.

The endoplasmic reticulum manufactures, processes, and transports a wide variety of biochemical compounds for use inside and outside of the cell. Consequently, many of the proteins found in the cisternal space of the endoplasmic reticulum lumen are there only transiently as they pass on their way to other locations. Other proteins, however, are targeted to constantly remain in the lumen and are known as endoplasmic reticulum resident proteins. These special proteins, which are necessary for the endoplasmic reticulum to carry out its normal functions, contain a specialized retention signal consisting of a specific sequence of amino acids that enables them to be retained by the organelle. An example of an important endoplasmic reticulum resident protein is the chaperone protein known as BiP (formally: the chaperone immunoglobulin-binding protein), which identifies other proteins that have been improperly built or processed and keeps them from being sent to their final destinations.

There are two basic kinds of endoplasmic reticulum morphologies: rough and smooth. The surface of rough endoplasmic reticulum is covered with ribosomes, giving it a bumpy appearance when viewed through the microscope. This type of endoplasmic reticulum is involved mainly with the production and processing of proteins that will be exported, or secreted, from the cell. The ribosomes assemble amino acids into protein units, which are transported into the rough endoplasmic reticulum for further processing. These proteins may be either transmembrane proteins, which become embedded in the membrane of the endoplasmic reticulum, or water-soluble proteins, which are able to pass completely through the membrane into the lumen. Those that reach the inside of the endoplasmic reticulum are folded into the correct three-dimensional conformation, as a flattened cardboard box might be opened up and folded into its proper shape in order to become a useful container. Chemicals, such as carbohydrates or sugars, are added, then the endoplasmic reticulum either transports the completed proteins to areas of the cell where they are needed, or they are sent to the Golgi apparatus for further processing and modification.

Most proteins exported from the endoplasmic reticulum exit the organelle in vesicles budded from the smooth portion, which has a more even appearance than rough endoplasmic reticulum when viewed through the electron microscope because of the lack of ribosomes. The smooth endoplasmic reticulum in most cells is much less extensive than the rough endoplasmic reticulum and is sometimes alternatively termed transitional. Smooth endoplasmic reticulum is chiefly involved, however, with the production of lipids (fats), building blocks for carbohydrate metabolism, and the detoxification of drugs and poisons. Therefore, in some specialized cells, such as those that are occupied chiefly in lipid and carbohydrate metabolism (brain and muscle) or detoxification (liver), the smooth endoplasmic reticulum is much more extensive and is crucial to cellular function. Smooth endoplasmic reticulum also plays a role in various cellular activities through its storage of calcium and involvement in calcium metabolism. In muscle cells, smooth endoplasmic reticulum releases calcium to trigger muscle contractions.

Accordingly, the nucleus has direct access to proteins (many of which are produced by the ribosomes upon its surface) and other materials present in the endoplasmic reticulum lumen, so that transport vesicles are not needed to obtain them. The close association between the endoplasmic reticulum and the nucleus also enables the organelles to share information in a very efficient manner. For instance, if the endoplasmic reticulum begins to undergo functional problems and unfolded proteins accumulate within the organelle, which can be extremely hazardous to the cell, the organelle quickly sends a signal to the nucleus (as well as to the cytoplasm). The nucleus responds by slowing ribosomal translation through a several-step process, thereby giving the endoplasmic reticulum extra time to catch up on its protein folding, thus maintaining cellular health.

Endosomes and Endocytosis

Endosomes are membrane-bound vesicles, formed via a complex family of processes collectively known as endocytosis, and found in the cytoplasm of virtually every animal cell. The basic mechanism of endocytosis is the reverse of what occurs during exocytosis or cellular secretion. It involves the invagination (folding inward) of a cell's plasma membrane to surround macromolecules or other matter diffusing through the extracellular fluid. The encircled foreign materials are then brought into the cell, and following a pinching-off of the membrane (termed budding), are released to the cytoplasm in a sac-like vesicle. The size of vesicles varies, and those larger than 100 nanometers in diameter are typically referred to as vacuoles.

Through receptor mediated endocytosis, active cells are able to take in significant amounts of particular molecules (ligands) that bind to receptor sites extending from the cytoplasmic membrane into the extracellular fluid surrounding the cell. These receptor sites are commonly grouped together along coated pits in the membrane, which are lined on their cytoplasmic surface with a bristle-like collection of coat proteins. The coat proteins are thought to play a role in enlarging the pit and forming a vesicle.

By means of pinocytosis, a cell is able to ingest droplets of liquid from the extracellular fluid. All solutes found in the droplets outside of the cell may become encased in the vesicles formed via this process, with those present in the greatest concentration in the extracellular fluid also becoming the most concentrated in the membranous sacs.

Pinocytic vesicles tend to be smaller than vesicles produced by other endocytic processes.

The final type of endocytosis, termed phagocytosis, is probably the most well-known manner in which a cell may import outside materials. In many school science labs, children observe amoebas under the microscope and watch the single-celled organisms eat by stretching out pseudopodia and encircling any food particles they find in their paths. This engulfment and subsequent packaging of the particles into vesicles, which are usually large enough to be correctly referred to as vacuoles, is phagocytosis. Though commonly associated with amoebas, phagocytosis is practiced by many organisms. In most multicellular animals, phagocytic cells chiefly function in bodily defense rather than as a means to gain nourishment. For example, leukocytes in the human body often phagocytose protozoa, bacteria, dead cells, and similar materials in order to help stave off infections or other problems.

Once freed into the cytoplasm, several small vesicles produced via endocytosis may come together to form a single entity. This endosome generally functions in one of two ways. Most commonly, endosomes transport their contents in a series of steps to a lysosome, which subsequently digests the materials. In other instances, however, endosomes are used by the cell to transport various substances between different portions of the external cell membrane. This latter function is particularly important among epithelial cells, such as those that compose the outer layer of the skin, because they exhibit polarity.

An endosome that is destined to transfer its contents to a lysosome generally goes through several changes along its way. In its initial form, when the structure is often referred to as an early endosome, the specialized vesicle contains a single compartment. Over time, however, chemical changes in the vesicle take place and the membrane surrounding the endosome folds in upon itself in a way that is similar to the invagination of the plasma membrane. In this case, however, the membrane is not pinched off. Consequently, a structure with multiple compartments, termed a multivesicular endosome, is formed. The multivesicular endosome is an intermediate structure in which further chemical changes, including a significant drop in pH level, take place as the vesicle develops into a late endosome.

STRUCTURE AND FUNCTION

Most cells are too small to be seen with the unaided eye. Cells were not discovered until after the invention of the microscope in the early seventeenth century. The cell theory consist of three principles:

1. All living things are composed of one or more cells.
2. Cells are the basic units of structure and function in an organism.
3. Cells come only from reproduction of existing cells.

Cell Diversity

Even cells within the same organism show enormous diversity in size, shape, and internal organization. Your body contains at least 200 different cell types.

Size

A few types of cells are large enough to be seen by the unaided eye. The female egg is the largest cell in the body, and can be seen without the aid of a microscope. Most cells are visible only with a microscope. Most cells are small for two reasons:

— Cells are limited in size by the ratio between their outer surface area and their volume. A small cell has more surface area than a large cell for a given volume of cytoplasm. This is important because the nutrients, oxygen, and other materials a cell requires must enter through it surface. As a cell grows larger at some point its surface area becomes too small to allow these materials to enter the cell quickly enough to meet the cell's need.

— The cell's nucleus (the brain) can only control a certain amount of living, active cytoplasm.

Shape

Cells come in a variety of shapes. Notice the neurons on the wall, the basic cell of our nervous system. This diversity of form reflects a diversity of function. Most cells have a specific shape. The shape of a cell depends on it's function. Cells of the nervous system that carry information from your toes to your brain are long and threadlike. Blood cells are shaped like round disk that can squeeze through tiny blood vessels.

Internal Organization

Cells contain a variety of internal structures called organelles. An organelle is a cell component that performs specific functions for the cell. Just as the organs of a multicellular organism carry out the organism's life functions, the organelles of a cell maintain the life of the cell. There are many different cells; however, there are certain features common to all, or most cells.

The entire cell is surrounded by a thin membrane, called the cell membrane. Inside the cell are a variety of organelles, most of which are surrounded by their own membrane. A large organelle near the center of the cell is the nucleus. It contains the cell's genetic information and controls the activities of the cell. The presence or absence of a nucleus is important for classifying cells.

Organisms whose cell contain a nucleus and other membrane-bound organelles are called eukaryotes. Organisms whose cells never contain (or lack) a nucleus and other membrane-bound organelles are called prokaryotes. Unicellular organisms such as bacteria and their relatives are prokaryotes. All other organisms are eukaryotes; plants, fish, mammals, insects and humans. The difference between prokaryotes and eukaryotes is such an important distinction that prokaryotes are placed in two kingdoms, separate from eukaryotes.

Eukaryotic Cell

The structures that make up a eukaryotic cell are determined by the specific functions carried out by the cell. Thus, there is no typical eukaryotic cell. Nevertheless, eukaryotic cells generally have three main components: a cell membrane, a nucleus, and other organelles.

Cell Membrane

A cell cannot survive if it is totally isolated from its environment. The cell membrane is a complex barrier separating the cell from it's external environment. This "selectively permeable" membrane regulates what passes into and out of the cell. All cells, from all organisms, are surrounded by a cell membrane. It is a thin layer of lipid and protein that separates the cell's content from the world around it. The cell membrane functions like a gate, controlling what enters and leaves the cell.

The cell membrane controls the ease with which substances pass into and out of the cell-some substances easily cross the membrane, while others cannot cross at all. For this reason, the cell membrane is said to be selectively permeable. Cell membranes are made mostly of phospholipid molecules.

Lipid is a simple form of fat. Phospholipids are a kind of lipid that consists of two fatty acids (tails), and phosphate group (heads). A phospholipid molecule has a polar "head" and two nonpolar "tails". Polar - the two ends of the phospholipid molecule have different properties in water. The phosphate head is hydrophilic meaning "water loving". Because of its hydrophilic nature, the head of a phospholipid will orient itself so that it is as close as possible to water molecules.

The lipid tails are hydrophobic meaning "water-fearing", the hydrophobic tails will tend to orient themselves away from water. When dropped in water, phospholipids line up on the surface with their phosphate heads sticking into the water and lipid tails pointing up from the surface. Cells are bathed in aqueous, or watery, environment. Since the inside of a cell is also an aqueous environment, both sides of the cell membrane are surrounded by water molecules. These water molecules cause the phospholipids of the cell membrane to form two layers.

Cell membranes consist of two phospholipid layers called a lipid bilayer. Heads face the watery fluids inside and outside the cell. Lipid tails are sandwich inside the bilayer. The cell membrane is constantly being formed and broken down in living cells.

Membrane proteins

A variety of protein molecules are embedded in the lipid bilayer. Some proteins are attached to the surface of the cell membrane, these are called peripheral proteins, and are located on both the internal and external surface. The proteins that are embedded in the lipid bilayer are called integral proteins. Some integral proteins extend across the entire cell membrane and are exposed to both the inside of the cell and the exterior environment. Others extend only to the inside or only to the exterior surface.

There are many kinds of proteins in membranes; they help to move material into and out of the cell. Some integral proteins form channels or pores through which certain substances can pass. Other proteins bind to a substance on one side of the membrane and carry it to the other side of the membrane. Integral proteins exposed to the cell's external environment often have carbohydrates attached to them serve as identification badges that allow cells to recognize each other and may act as site where viruses or chemical messengers such as hormones can attach.

Fluid mosaic model of cell membranes

Membranes are fluid and have the consistency of vegetable oil. The lipids and proteins of the cell membrane are always in motion. Phospholipids are able to drift across the membrane, changing places with their neighbor. Proteins in and on the membrane form patterns, or mosaics. Because the membrane is fluid with a mosaic of proteins, scientists call the modern view of membrane structure the fluid mosaic model. The pattern, or "mosaic" of lipids and proteins in the cell membrane is constantly changing.

Cytoplasm

Everything between the cell membrane and the nucleus is the cell's cytoplasm. Cytoplasm consists of two main components: cytosol and organelles. Cytosol is a jellylike mixture that consists mostly of water, along with proteins, carbohydrates, salts, minerals and organic molecules. Suspended in the cytosol are tiny organelles (organs). Organelles are structures that work like miniature organs, they carry out specific functions in the cell. The organelles plus the cytosol makes up the cytoplasm. In eukaryotic cells, most organelles are surrounded by a membrane. Prokaryotic cells have no membrane-bound organelles.

Mitochondria

Mitochondria are found scattered throughout the cytosol, and are relatively large organelles. Mitochondria are the sites of chemical reactions that transfer energy from organic compounds to ATP. Energy contain in food is released. Converted to ATP. ATP is the molecule that most cells use as their main energy currency. It is the "powerhouse" of the cell.

Mitochondria are usually more numerous in cells that have a high energy requirement - your muscle cells contain a large number of mitochondria. Mitochondria is surrounded by two membranes. The smooth outer membrane serves as a boundary between the mitochondria and the cytosol. The inner membrane has many long folds, known as cristae (kris-tee). The cristae greatly increases the surface area of the inner membrane, providing more space for the chemical reactions to occur. Mitochondria have their own dna, and new mitochondria arise only when existing ones grow and divide. ATP production is called cellular respiration.

Ribosomes

Unlike most other organelles, ribosomes are not surrounded by a membrane. Ribosomes are the site of protein synthesis (production or construction) in a cell. They are most numerous organelles in almost all cells. Some are free in the cytoplasm; others line the membranes of rough endoplasmic reticulum.

Endoplasmic reticulum (ER)

The ER is a system of membranous tubules and sacs. The ER functions primarily as an intracellular highway, a path along which molecules move from one part of the cell to another. The amount of ER inside a cell fluctuates, depending on the cell's activity. Poisons, waste, and other toxic chemicals are made harmless. ER is an extensive network of membranes that connect the nuclear envelope to the cell membrane. It transports the materials through cell. It can be rough or smooth.

Rough ER is studded with ribosomes and processes proteins to be exported from the cell. Smooth ER is not covered with ribosomes and processes lipids and carbohydrates. The smooth ER is involved in the synthesis of steroids in gland cells, the regulation of calcium levels in muscle cells, and the breakdown of toxic substances by liver cells.

Golgi Apparatus

The golgi apparatus is the processing, packaging and secreting organelle of the cell. It is a system of membranes. Made of flattened sac like structures called Cisternae. It works closely with the er, the golgi apparatus modifies proteins for export by the cell.

Lysosomes

Lysosomes are small spherical organelles that enclose hydrolytic enzymes within a single membrane. Lysosomes are the site of food digestion in the cell. Lysosomes are formed from pieces of the golgi apparatus that break off. Lysosomes are common in the cells of animals, fungi, and protists, but rare in plant cells.

Cytoskeleton

Just as your body depends on your skeleton to maintain its shape and size, so a cell needs structures to maintain its shape and size. In animal cells, an internal framework called cytoskeleton maintains the shape of the cell. The cytoskeleton maintains the three-dimensional structure of the cell, participates in the movement of organelles within the cytosol, and helps the cell move. It is a network of long protein strands located in the cytosol, that are not surrounded by a membrane. The cytoskeleton consists of two types of structures: microfilaments and microtubules.

Microfilaments

Microfilaments are not hallow and have a structure that resembles rope made of two twisted chains of protein called actin. It can be contract, causing movement. Muscle cells have many microfilaments.

Microtubules

Microtubules are hallow tubes like plumbing pipes. They are the largest strands of the cytoskeleton. They are made of a protein called tubulin. It have three functions:

1. To maintain the shape of the cell.
2. To serve as tracks for organelles to move along within the cell.
3. When the cell is about to divide, bundles of microtubules known as spindle fibers come together and extend across the cell to assist in the movement of chromosomes during cell division.

Cilia and Flagella

Cilia and flagella are hairlike organelles that extend from the surface of the cell, where they assist in movement. Microtubules are sometimes bundled into structures called cilia and flagella. Cilia are short hairlike projections. Flagella are long whiplike projections. The cilia and flagella of all eukaryotes consist of one pair of microtubules surrounded by nine more pairs. Cilia are often numerous. Flagella are often singular.

Unicellular organisms such as paramecium and euglena use cilia and flagella to move through water. Sperm use flagella to swim to the egg. In humans, beating cilia line parts of the respiratory system, moving dust particles and bacteria away from the lungs.

Nucleus

The nucleus is often the most prominent structure within

a eukaryotic cell. It maintains its shape with the help of a protein skeleton known as the nuclear matrix. The nucleus is the control center (brain) of the cell. Most cells have a single nucleus some cells have more than one. The nucleus is surrounded by a double layer membrane called the nuclear envelope. The nuclear envelope is covered with many small pores through which proteins and chemical messages from the nucleus can pass.

The nucleus contains DNA, the hereditary material of cells. The DNA is in the form of a long strand called chromatin. During cell division, chromatin strands coil and condenses into thick structures called chromosomes. The chromosomes in the nucleus contain coded "blueprints" that control all cellular activity. Most nuclei contain at least one nucleolus (plural, nucleoli). The nucleolus makes (syntheisizes) ribosomes, which in turn, build proteins. When a cell prepares to reproduce, the nucleolus disappears.

Plant Cells

Most of the organelles and other parts of the cell are common in all eukaryotic cells. Cell from different organisms have even greater difference in structure. Plant cells have three additional structures not found in animals cells - cell walls, vacuoles, and plastids that are extremely important to plant function. In addition to their unique structures, plant cells have: mitochondria, ribosomes, and the other organelles.

Cell wall

One of the most important differences between plant and animal cells is the presence of a cell wall in plant cells. Fungi such as mushrooms and yeast also have cell walls. Cell walls of fungi are made of chitin. A cell wall does not replace the cell membrane; cells with walls also have a cell membrane. Plant cells are covered by a rigid cell wall that lies outside the cell membrane. The rigidity of cell walls helps support and protect the plant. Cell walls of plants contain polysaccharide (long chains) cellulose a complex carbohydrate. Cell walls are of two types:

1. *Primary cell wall*: while a plant cell is being formed, a primary cell wall develops just outside the cell membrane. As the cell expands in length, cellulose and other molecules are added, enlarging the cell wall. When the cell reaches full size, a secondary cell wall may form.
2. *Secondary cell wall*: the secondary cell walls forms between the primary cell wall and the cell membrane. The secondary cell wall is tough and woody, in fact the secondary cell wall is what we call wood. One a secondary cell wall forms, a plant cell can grow no further. The cells are dead.

Vacuoles

The second prominent structure in plant cells is the large vacuole. The vacuole is a large membrane-bound sac that takes up a large amount of space in most plant cells. It serves as a storage area, and may contain stored proteins, ions, waste, or other cell products. Vacuoles of some plants contain poison that discourages animals from eating the plant's leaves. Cells of animals and other organisms also may contain vacuoles, but they are much smaller and are usually involved in food digestion.

Plastids

A third distinguishing feature of plant cells is the presence of structures called plastidsthat make or store food. A common kind of plastid is the chloroplast, an organelle that converts sunlight, carbon dioxide, and water into sugars. This process is called photosynthesis. Each chloroplast encloses a system of flattened, membranous sacs called thylakoids. It is in the thylakoids that photosynthesis occurs. Chloroplasts are green because they contain chlorophyll, a pigment that absorbs energy in sunlight. They are found only in algae, such as seaweed, and in green plants. Other plastids store reddish-orange pigments that color fruits, vegetables, flowers, and autumn leaves.

MULTICELLULAR ORGANIZATION

In a unicellular organism, one cell carries out all of the functions of life. In contrast, most cells in a multicellular organism are specialized to perform one or a few functions. Because of cell specialization, the cells of multicellular organisms depend on other cells in the organism for their survival.

Tissue, Organs, and Organ Systems

In most multicellular organisms have:

— *Cellular level*: the smallest unit of life capable of carrying out all the functions of living things.

— *Tissue level*: a group of cells that performs a specific function in an organism form the tissue.

— *Organ level*: several different types of tissue that function together for a specific purpose form an organ.

— *Organ system level*: several organs working together to perform a function make up an organ system. The different organ systems in a multicellular organism interact to carry out the processes of life

Plants also have tissue and organs, although they are arranged somewhat differently from those of animals.

1. Dermal tissue system forms the outer layer of a plant.
2. Ground tissue system makes up the bulk of roots and stems
3. Vascular tissue transports water and food throughout the plant.

4. The four plant organs are roots, stems, leaves and flowers.

Colonial organizations

A colonial organization is a collection of genetically identical cells that live together in a closely connected group. Many of the cells of the colony carry out specific functions that benefit the whole colony. Colonial organisms appear to straddle the border between a collection of unicellular organisms and a true multicellular organism. They lack tissues and organs, but do exhibit the principle of cell specialization.

Cell Signaling

Information can come in a variety of forms, and communication frequently involves converting information signals from one form to another. The signals that pass between cells are far simpler than the sorts of messages that humans ordinarily exchange. In a typical communication between cells, the signaling cell produces a particular type of signal molecule that is detected by the target cell.

The target cells possess receptor proteins that recognize and respond specifically to the signal molecule. Signal transduction begins when the receptor protein on the target cell receives an incoming extracellular signal and converts it to the intracellular signals that direct cell behavior. Single cells and cells in multicellular organisms use hundreds of kinds of extracellular molecules to send signals to one another-proteins, peptides, amino acids, nucleotides, steroids, fatty acid derivatives, and even dissolved gases-but they rely on only a handful of basic styles of communication for getting the message across. In multicellular organisms, the most "public" style of communication involves broadcasting the signal throughout the whole body by secreting it into the bloodstream (in an animal) or the sap (in a plant).

Signal molecules used in this way are called hormones, and in animals, the cells that produce hormones are called endocrine cells. For example, part of the pancreas is an endocrine gland that produces the hormone insulin, which regulates glucose uptake in cells all over the body. Somewhat less public is the process known as paracrine signaling. In this case, rather than entering the bloodstream, the signal molecules diffuse locally through the extracellular medium, remaining in the neighborhood of the cell that secretes them. Thus they act as local mediators on nearby cells.

Many of the signal molecules that regulate inflammation at the site of an infection or control cell proliferation in a healing wound function in this way. Neuronal signaling constitutes a third form of cell communication. Like endocrine cells, neurons can deliver messages across long distances. In the case of neuronal signaling, however, a message is not broadcast widely but is delivered quickly and specifically to individual target cells through private lines. The axon of a neuron terminates at specialized junctions (synapses) on target cells that can lie far from the neuronal cell body. The axons that connect a person's spinal cord and big toe, for example, can be more than 1 m in length.

When activated by signals from the environment or from other nerve cells, a neuron sends electrical impulses racing along its axon at speeds of up to 100 m/sec. On reaching the axon terminal, these electrical signals are converted into a chemical form: each electrical impulse stimulates the nerve terminal to release a pulse of an extracellular chemical signal called a neurotransmitter. These neurotransmitters then diffuse across the narrow (< 100 nm) gap between the axon-terminal membrane and the membrane of the target cell in less than 1 msec. A fourth style of signal-mediated cell-cell communication-the most intimate and short-range of all-does not require the release of a secreted molecule.

Instead, the cells make direct contact through signaling molecules lodged in their plasma membranes. The message is delivered when a signal molecule anchored in the plasma membrane of the signaling cell binds to a receptor molecule embedded in the plasma membrane of the target cell. In embryonic development, for example, such contact-dependent signaling plays an important part in tissues in which adjacent cells that are initially similar are destined to become specialized in different ways. To relate these different signaling styles, imagine trying to advertise a potentially stimulating lecture-or a concert or football game.

An endocrine signal would be akin to broadcasting the information over a radio station. A flyer posted on select notice boards would be the equivalent of a localized paracrine signal. Neuronal signals-long-distance but personal-would be similar to a phone call or an e-mail, and contact-dependent signaling would be like a good, old-fashioned face-toface conversation.

A typical cell in a multicellular organism is exposed to hundreds of different signal molecules in its environment. These may be free in the extracellular fluid, embedded in the extracellular matrix in which cells rest, or bound to the surfaces of neighboring cells. Each cell must respond selectively to this mixture of signals, disregarding some and reacting to others, according to the cells' specialized function. Whether a cell responds to a signal molecule depends first of all on whether it possesses a receptor for that signal. Without the appropriate receptor, a cell will be deaf to the signal and will not react. By producing only a limited set of receptors out of the thousands that are possible, the cell restricts the range of signals that can affect it. But this limited range of signals can still be used to control the behavior of the cell in complex ways. The complexity is of two sorts.

First, one signal, binding to one type of receptor protein, can cause a multitude of effects in the target cell: it can alter the cell's shape, movement, metabolism, and gene expression. The signal from a cell-surface receptor is generally conveyed into the cell interior via a set of interacting molecular mediators that are capable of producing widespread effects in the cell. Furthermore, this intracellular relay system and the intracellular targets on which it acts vary from one type of specialized cell to another, so that different types of cells respond to the same signal in different ways. For example, when a heart muscle cell is exposed to the neurotransmitter acetylcholine, the rate and force of its contractions decrease, but when a salivary gland is exposed to the same signal, it secretes components of saliva. These responses occur rapidly-within seconds to minutes-because the signal affects the activity of proteins and other molecules that are already present inside the cells.

The second kind of complexity arises because a typical cell possesses a whole collection of different receptors-tens to hundreds of thousands of receptors of a few dozen types. Such variety makes the cell simultaneously sensitive to many extracellular signals. These signals, by acting together, can evoke responses that are more than just the sum of the effects that each signal would evoke on its own. The intracellular relay systems for the different signals interact, so that the presence of one signal modifies the responses to another. Thus one combination of signals may simply enable a cell to survive; another may drive it to differentiate in some specialized way; and another may cause it to divide.

In the absence of any signals, most animal cells are programmed to kill themselves. Because the execution of such a complex programme often requires the synthesis of new proteins, it may take the cell hours to fully respond to the incoming signals. Overall, the integration of extracellular cues allows a relatively small number of signal molecules, used in different combinations, to exert subtle and complex control over cell behavior.

Signal reception begins at the point where a signal originating outside the target cell encounters a target molecule belonging to the cell itself. In virtually every case, the target molecule is a receptor protein (also called a receptor), and each receptor is usually activated by only one type of signal. The receptor protein performs the primary transduction step: it receives an external signal and generates a new intracellular signal in response.

As a rule, this is only the first event in a chain of intracellular signal transduction processes. In this game of molecular tag, the message is passed from one intracellular signaling molecule to another, each activating or generating the next signaling molecule in line, until, say, a metabolic enzyme is kicked into action, a gene is switched on, or the cytoskeleton is tweaked into a new configuration. This final outcome is called the response of the cell. These relay chains, or signaling cascades, of intracellular signaling molecules have several crucial functions:

1. They transform, or transduce, the signal into a molecular form suitable for passing the signal along or stimulating a response.
2. They relay the signal from the point in the cell at which it is received to the point at which the response is produced.
3. In many cases, signaling cascades also amplify the signal received, making it stronger, so that a few extracellular signal molecules are enough to evoke a large intracellular response.

4. The signaling cascades can also distribute the signal so as to influence several processes in parallel: at any step in the pathway, the signal can diverge and be relayed to a number of different intracellular targets, creating branches in the information flow diagram and evoking a complex response.
5. Each step in this signaling cascade is open to modulation by other factors, including other external signals, so that the effects of the signal can be tailored to the conditions prevailing inside or outside the cell.

Most signaling pathways trace a long and branching route, and enlist many molecular players, as they relay information from receptors at the cell surface to appropriate machinery in the cell's interior.

Extracellular Signal Molecules

Extracellular signal molecules in general fall into two classes. The first and largest class of signals consists of molecules that are too large or too hydrophilic to cross the plasma membrane of the target cell. They rely on receptors on the surface of the target cell to relay their message across the membrane. The second, and smaller, class of signals consists of molecules that are small enough or hydrophobic enough to slip easily through the plasma membrane. Once inside, these signal molecules either activate intracellular enzymes or bind to intracellular receptor proteins that regulate gene expression.

For an extracellular signal to alter a cell within a few seconds or minutes, direct activation of an enzyme is an effective strategy. Nitric oxide (NO) is just such a signal. This dissolved gas diffuses readily out of the cell that generates it and enters neighboring cells. NO is made from the amino acid arginine and operates as a local mediator in many tissues. The gas acts only locally because it is quickly converted to nitrates and nitrites (with a half-life of about

5-10 seconds) by reaction with oxygen and water outside cells. Endothelial cells-the flattened cells that line every blood vessel-release NO in response to stimulation by nerve endings. This NO signal causes smooth muscle cells in the vessel wall to relax, allowing the vessel to dilate, so that blood flows through it more freely.

The effect of NO on blood vessels accounts for the action of nitroglycerine, which has been used for almost 100 years to treat patients with angina (pain caused by inadequate blood flow to the heart muscle). In the body, nitroglycerine is converted to NO, which relaxes coronary blood vessels and increases blood flow to the heart. Many nerve cells also use NO to signal neighboring cells: NO released by nerve terminals in the penis, for instance, triggers the local blood-vessel dilation that is responsible for penile erection. Inside many target cells, NO binds to the enzyme guanylyl cyclase, stimulating the formation of cyclic GMP from the nucleotide GTP.

Cyclic GMP itself is a small intracellular signaling molecule that forms the next link in the signaling chain that leads to the cell's ultimate response. The impotence drug Viagra enhances penile erection by blocking the degradation of cyclic GMP, prolonging the NO signal. Cyclic GMP is very similar in its structure and mechanism of action to cyclic AMP.

Gases such as NO are not the only signal molecules that can cross the plasma membrane. Hydrophobic signal molecules such as the steroid hormones-including cortisol, estradiol, and testosterone-and the thyroid hormones such as thyroxine all pass through the plasma membrane of the target cell. Instead of activating intracellular enzymes, however, they bind to receptor proteins located either in the cytosol or in the nucleus. These hormone receptors are proteins capable of regulating gene transcription, but they are typically present in an inactive form in unstimulated cells.

When a hormone binds, the receptor protein undergoes a large conformational change that activates the protein, allowing it to promote or inhibit the transcription of a selected set of genes. Each hormone binds a different receptor protein, and each receptor acts at a different set of regulatory sites in DNA. Because the hormones regulate different sets of genes, they evoke a variety of physiological responses. Steroid hormone receptors play an essential role in human physiology, as illustrated by the dramatic consequences of a lack of the receptor for testosterone in humans. The male sex hormone testosterone shapes the formation of the external genitalia and influences brain development in the fetus; at puberty, it triggers the development of male secondary sexual characteristics.

Some very rare individuals are genetically male (that is, they have both an X and a Y chromosome) but lack the testosterone receptor as a result of a mutation in the corresponding gene; thus they make the hormone, but their cells cannot respond to it. As a result, these individuals develop the appearance and behavior of females, which is the pathway of sexual and brain development that would occur if no male or female hormones were produced. This demonstrates the key role of the testosterone receptor in sexual development, and also shows that the receptor is required not just in one cell type to mediate one effect of testosterone, but in many cell types to help produce the whole range of features that distinguish men from women.

Cell-Surface Receptors

In contrast to NO and the steroid and thyroid hormones, the vast majority of signal molecules are too large or hydrophilic to cross the plasma membrane of the target cell. These proteins, peptides, and other bulky, water-soluble molecules bind to receptor proteins that span the plasma membrane. The transmembrane receptors detect a signal on the outside and relay the message, in a new form, across the membrane into the interior of the cell.

Most cell-surface receptor proteins belong to one of three large families: ion-channel-linked receptors, G-protein-linked receptors, or enzyme-linked receptors. These families differ in the nature of the intra-cellular signal that they generate when the extracellular signal molecule binds to them. For ion-channel-linked receptors, the resulting signal is a flow of ions across the membrane, which produces an electrical current. G-protein-linked receptors activate a class of membrane-bound protein (a trimeric GTP-binding protein or G protein), which is then released to migrate in the plane of the plasma membrane, initiating a cascade of other effects.

And enzyme-linked receptors, when activated, act as enzymes or are associated with enzymes inside the cell. Switching on this enzymatic activity then generates a host of additional signals, including small molecules that are released into the cytosol. The number of different types of receptors in these three classes is even greater than the number of extracellular signals that act on them, because for many extracellular signal molecules there is more than one type of receptor. The neurotransmitter acetylcholine, for example, acts on skeletal muscle cells via an ion-channel-linked receptor, whereas in heart muscle cells it acts through a G-protein-linked receptor. These two types of receptors generate different intracellular signals, and thus enable the two types of muscle cells to react to acetylcholine in different ways, increasing contraction in skeletal muscle and decreasing the frequency of contractions in heart.

The multitude of different cell-surface receptors that the body requires for its own signaling purposes are also targets for many foreign substances that interfere with our physiology and sensations, from heroin and nicotine to tranquilizers and chili peppers. These substances either mimic the natural ligand for a receptor, occupying the normal ligand-binding site, or bind to the receptor at some other site, blocking or overstimulating the receptor's natural activity. Many drugs and poisons act in this way, and a large part of the pharmaceutical industry is devoted to the search for substances that will exert a precisely defined effect by binding to a specific type of cell-surface receptor.

Ion-channel-linked Receptors

Of all the cell-surface receptor types, ion-channel-linked receptors (also known as transmitter-gated ion channels) function in the simplest and most direct way. These receptors are responsible for the rapid transmission of signals across synapses in the nervous system. They transduce a chemical signal, in the form of a pulse of neurotransmitter delivered to the outside of the target cell, directly into an electrical signal, in the form of a change in voltage across the target cell's plasma membrane. When the neurotransmitter binds, this type of receptor alters its conformation so as to open or close a channel for the flow of specific types of ions-such as Na^+, K^+, Ca^{2+}, or Cl^-—across the membrane.

Driven by an electrochemical gradient, the ions rush into or out of the cell, creating a change in the membrane potential within a millisecond or so. This change in potential may trigger a nerve impulse, or alter the ability of other signals to do so. The opening of Ca^{2+} channels has special effects, as changes in the intracellular Ca^{2+} concentration can profoundly alter the activities of many enzymes. Whereas ion-channel-linked receptors are a specialty of the nervous system and of other electrically excitable cells such as muscle, Gprotein -linked receptors and enzyme-linked receptors are used by practically every cell type of the body.

REFERENCES

Alberts B. et al., *Molecular Biology of the Cell,* 2d ed. 1989.
Lackie J. M. and J. A. Dowe, ed., *The Dictionary of Cell Biology,* 1989.
Prescott D. M., *Cells,* 1988.
Thomas L., *The Lives of a Cell,* 1974.

2

Cell Division

Cell division is a process by which a cell, called the parent cell, divides into two cells, called *daughter cells*. Cell division is usually a small segment of a larger cell cycle. In meiosis however, a cell is permanently transformed and cannot divide again.

For simple unicellular organisms such as the Amoeba, one cell division reproduces an entire organism. On a larger scale, cell division can create progeny from multicellular organisms, such as plants that grow from cuttings. Cell division also enables sexually reproducing organisms to develop from the one-celled zygote, which itself was produced by cell division from gametes. And after growth, cell division allows for continual renewal and repair of the organism.

The primary concern of cell division is the maintenance of the original cell's genome. Before division can occur, the genomic information which is stored in chromosomes must be replicated, and the duplicated genome separated cleanly between cells. A great deal of cellular infrastructure is involved in keeping genomic information consistent between "generations".

CELL DIVISION AND DNA RECOMBINATION

The cell cycle, or cell-division cycle, is the series of events that take place in a eukaryotic cell leading to its replication.

These events can be divided in two brief periods: interphase—during which the cell grows, accumulating nutrients needed for mitosis and duplicating its DNA—and the mitotic (M) phase, during which the cell splits itself into two distinct cells, often called "daughter cells". The cell-division cycle is a vital process by which a single-celled fertilised egg develops into a mature organism, as well as the process by which hair, skin, blood cells, and some internal organs are renewed.

The cell cycle consists of four distinct phases: G_1 phase, S phase, G_2 phase (collectively known as interphase) and M phase. M phase is itself composed of two tightly coupled processes: mitosis, in which the cell's chromosomes are divided between the two daughter cells, and cytokinesis, in which the cell's cytoplasm divides forming distinct cells. Activation of each phase is dependent on the proper progression and completion of the previous one. Cells that have temporarily or reversibly stopped dividing are said to have entered a state of quiescence called G_0 phase.

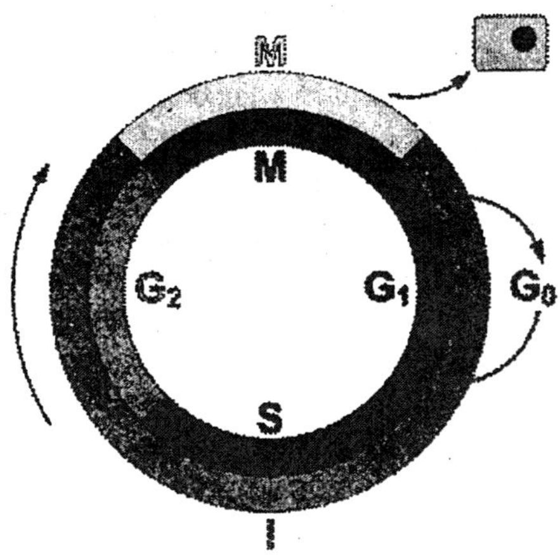

Figure 1. The cell cycle.

Without counting the G_0 phase, a cell cycle takes 12-24 hours for most mammalian cells, and only 20-30 minutes for E. coli cells.

CDK AND CYCLINS

The progression of cell cycle is catalysed by cyclin-dependent kinase (CDK) which, as the name suggests, is activated by a special class of proteins called cyclins. In mammals, different cyclins are designated as A, B, C, D and so on. In budding yeast, nine cyclins have been identified: Cln1 to Cln3 and Clb1 to Clb6. The proteins of the CDK superfamily in mammals is denoted by Cdk followed by a number. However, in yeast, they are denoted by Cdc (cell division control) and a number. Among them, Cdc2 was first identified in fission yeast. It is equivalent to Cdc28 in budding yeast, and to Cdk1 in mammals. Cdk1 is often referred to as Cdc2.

In budding yeast, the cell cycle is initiated by the binding of Cln1 or Cln2 to Cdc28. Then, Clb5 or Clb6 binds to Cdc28, and pushes the cycle forward. Clb1 and Clb2 are involved in the final phase. In mammals, the cell cycle is initiated by the binding of Cyclin D (CycD) to Cdk4 or Cdk6. The final phase is catalysed by the binding between Cyclin B (CycB) and Cdk1. the CycB-Cdk1 complex is known as MPF, which originally stands for maturation promoting factor, but now also stands for mitosis promoting factor.

Initiation and Termination Mechanism of S phase

DNA replication occurs in the S phase of the cell cycle. DNA replication is triggered by the expression of all required proteins. In mammals, the expression of these necessary proteins is activated by the transcription factor E2F which, in turn, is regulated by the protein pRB (the protein product of the retinoblastoma gene). When pRB binds to E2F, it will inactivate the transcriptional function of E2F, thereby inhibiting DNA replication. With such a critical role, pRB is known as the "master brake of cell division".

During G_0 or early G_1 phase, E2F is inactivated by pRB. In late G_1 phase, the Cyclin D-Cdk4 complex phosphorylates a specific set of sites on pRB, resulting in the release of E2F, which then leads to the production of proteins needed for DNA replication. E2F also stimulates the production of Cyclin E, cyclin A and E2F itself. The Cyclin E-Cdk2 complex can also phosphorylate pRB, thus accelerating the cell cycle advance into the S phase. On the other hand, the Cyclin A-Cdk2 complex can phosphorylate E2F, abolishing its transcriptional function. Consequently, DNA replication is terminated.

It is interesting to note that the Cyclin E-Cdk2 complex cannot phosphorylate E2F. Otherwise, the DNA replication would be terminated prematurely.

p53

p53 is a tumor suppressor protein, also known as "Guardian of the Genome". It plays an important role in cell cycle control and apoptosis. Defective p53 could allow abnormal cells to proliferate, resulting in cancer. As many as 50% of all human tumors contain p53 mutants. In normal cells, the p53 protein level is low. DNA damage and other stress signals may trigger the increase of p53 proteins, which have three major functions: growth arrest, DNA repair and apoptosis (cell death).

The growth arrest stops the progression of cell cycle, preventing replication of damaged DNA. During the growth arrest, p53 may activate the transcription of proteins involved in DNA repair. Apoptosis is the "last resort" to avoid proliferation of cells containing abnormal DNA. The cellular concentration of p53 must be tightly regulated. While it can suppress tumors, high level of p53 may accelerate the aging process by excessive apoptosis.

The major regulator of p53 is Mdm2, which can trigger the degradation of p53 by the ubiquitin system.

Target Genes

p53 is a transcriptional activator, regulating the expression of Mdm2 (for its own regulation) and the genes involved in growth arrest, DNA repair and apoptosis. Some important examples are given below.

— Growth arrest: *p21, Gadd45, and 14-3-3σ.*

— DNA repair: p53R2.

— Apoptosis: Bax, Apaf-1, PUMA and NoxA.

Regulation of p53

p53 is mainly regulated by Mdm2.

a) Expression of Mdm2 is activated by p53.

b) Binding of p53 by Mdm2 can trigger the degradation of p53 via the ubiquitin system.

c) Phosphorylation of p53 at Ser15, Thr18 or Ser20 will disrupt its binding with Mdm2. In normal cells, these three residues are not phosphorylated, and p53 is maintained at low level by Mdm2.

d) DNA damage may activate protein kinase (such as ATM, DNA-PK, or CHK2) to phosphorylate p53 at one of these three residues, thereby increasing p53 level. Since Mdm2 expression is activated by p53, the increase of p53 also increases Mdm2, but they have no effect while p53 is phosphorylated. After the DNA damage is repaired, the ATM kinase is no longer active. p53 will be quickly dephosphorylated and destroyed by the accumulated Mdm2.

Roles of p53

p53 is also directly involved in DNA repair. One of its transcriptional target gene, p53R2, encodes ribonucleotide reductase, which is important for both DNA replication and repair. p53 also interacts directly with AP endonuclease and DNA polymerase which are involved in base excision repair.

a) The cell cycle progression into the S phase requires the enzyme Cdk2, which can be inhibited by p21. The progression into the M phase requires Cdc2 which can be inhibited by p21, GADD45 or 14-3-3s. p53 regulates the expression of these inhibitory proteins to induce growth arrest.

b) Apoptosis can be induced by the binding of Caspase 9 to cytochrome c and Apaf1. p53 may activate the expression of Apaf1 and Bax. The latter can then stimulate the release of cytochrome c from mitochondria.

DNA replication occurs in the S phase of the cell cycle. DNA replication is triggered by the expression of all required proteins. In mammals, the expression of these necessary proteins is activated by the transcription factor E2F which, in turn, is regulated by the protein pRB (the protein product of the retinoblastoma gene). When pRB binds to E2F, it will inactivate the transcriptional function of E2F, thereby inhibiting DNA replication. With such a critical role, pRB is known as the "master brake of cell division".

MITOSIS

Mitosis is a process of cell division resulting in two genetically equivalent daughter cells. The period between mitoses is called the interphase which covers the G_1, S, and G_2 phases of the cell cycle. During the G_2 phase, DNA has been replicated and the chromosome number is 4n. In the interphase, chromosomes are not visible by light microscopy because the chromatin diffuses throughout the nucleus.

— *Prophase*: The chromatin condenses into chromosomes. Two centrioles move toward the opposite ends (poles) of the cell. Microtubules radiate from each centriole, forming a structure called spindle. By the end of prophase, the nuclear envelope breaks down into small vesicles.

— *Metaphase*: The chromosomes align at the equatorial plane of the cell, as directed by the microtubules attached to the kinetochore (located at the centromere of each chromosome).

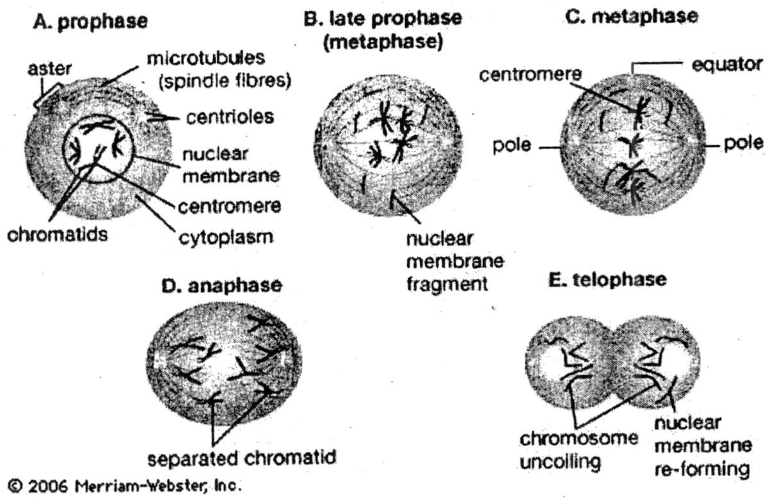

Figure 2. Stages of mitosis.

— *Anaphase*: Sister chromatids separate and move apart. The separation happens simultaneously for the entire set of chromosomes. The microtubules attached to the kinetochore also play a critical role in this process.

— *Telophase*: The chromosomes begin to uncoil and become less condensed. The spindle disappears while the nuclear envelope reappears. The cell elongates and eventually divides into two daughter cells. The process of cell division is called cytokinesis.

MPF and the Initiation

The morphological change in early mitosis is mainly due to the phosphorylation of related proteins catalysed by MPF, which is actually the Cyclin B/Cdk1 complex. The activated MPF may phosphorylate histone protein H1 to condense chromatin. It can also phosphorylate nuclear lamins to break nuclear envelope. The microtubular structure is closely related to MAPs (microtubule-associated proteins) which can also be phosphorylated by MPF, resulting in the formation of spindle. Cdc2 is a synonym of Cdk1. Y and T represent the specific Tyrosine and Threonine residue, respectively. MPF is active only when it is monophosphorylated at the specific Threonine residue.

Termination Mechanism

To exit mitosis, Cyclin A and Cyclin B must be destroyed. This is triggered by MPF which can phosphorylate DBRP (destruction box recognition proteins) to degrade these cyclins. During mitosis, the activity of MPF keeps rising until it phosphorylates DBRP to degrade Cyclin B, thereby inhibiting MPF. The morphological changes in early mitosis are mainly due to the phosphorylation of related proteins by MPF. After MPF is inhibited in late mitosis, those morphological changes are reversed: chromatin less condensed, nuclear envelope reformed and spindle disappears.

MEIOSIS

Meiosis is a special type of cell division resulting in four genetically nonequivalent daughter cells, each containing half the number of chromosomes of the parent cell. Germ cells (eggs and sperm) are produced by meiosis. The meiosis involves two consecutive cell divisions. Each division is composed of four stages as in mitosis. However, the first division differs in two important aspects:

a) In the metaphase I, each pair of sister chromatids aligns with its homologous pair side by side, forming two lines of sister chromatids at the equatorial plane (there is only one line in mitosis). Such alignment is called synapsis.

b) In the subsequent anaphase I, each pair of sister chromatids moves away from its homologous pair, but the sister chromatids are not separated.

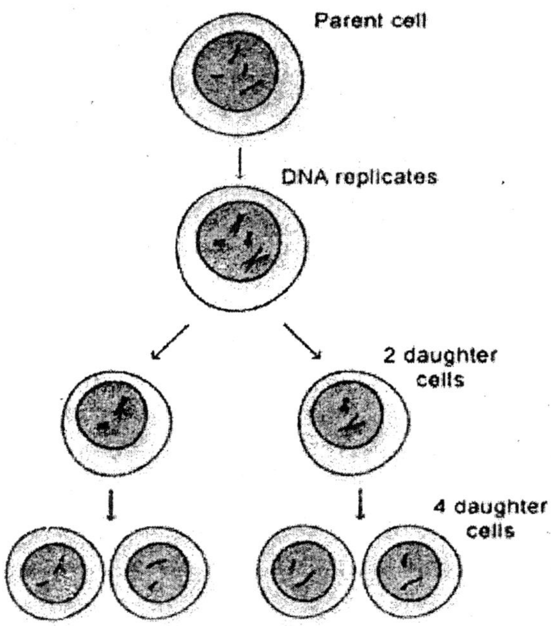

Figure 3. Meosis

After the first division, the chromosome number in each daughter cell becomes 2n. The process of the second division is the same as in mitosis except that there is no DNA replication during the interphase. As a result, the produced cell is a haploid (1n). During synapsis, DNA recombination may occur between homologous pairs of sister chromatids.

Independent Assortment

In the metaphase I of meiosis, two pairs of homologous sister chromatids align side by side. One pair (the maternal homolog) comes from the egg and another pair (the maternal homolog) comes from the sperm. Subsequently, the two homologous pairs are separated into two different daughter cells. Since they are randomly aligned, which homolog will enter which daughter cell is random. As a result, each daughter cell will contain some maternal homologs and some paternal homologs. For human cells, there are a total of 8.4 million (= 2^{23}) possible combinations.

Sperm cells (1n) come from the meiotic division of primary spermatocytes (2n). In their 23 chromosomes, some of them are inherited from the maternal homolog and others are from the paternal homolog.

DNA RECOMBINATION

DNA recombination refers to the process that a DNA segment moves from one DNA molecule to another DNA molecule. The following three types are most commonly observed.

— Homologous recombination

— Site-specific recombination

— Transpositional recombination

The homologous recombination often occurs during meiosis. Other types of recombination are not specifically related to cell division.

Homologous Recombination

Homologous recombination occurs between two homologous DNA molecules. It is also called DNA crossover. During meiosis, two homologous pairs of sister chromatids align side by side. The DNA crossover is very likely to occur. It could be as often as several times per meiosis.

- — Holliday model of DNA crossover
- — Gene conversion
- — Unequal crossover

Site-specific recombination

Site-specific recombination occurs at a specific DNA sequence. The first example was found in the integration between λ DNA and E. coli DNA. Both of them contain a sequence, 5'-TTTATAC-3', called the attachment site, which allows the two DNA molecules to attach together by base pairing.

Once attached, the enzyme integrase catalyzes two single strand breaks as in the Holliday model. After a short branch migration, the integrase exerts a second strand cuts on two other strands. Resolution of two Holliday junctions completes the integration process.

Transpositional Recombination

Transpositional recombination is the process that a mobile element is inserted into a target DNA. It may occur by one of two mechanisms: (1) directly as DNA, (2) through RNA. The mobile elements that transpose through DNA are called transposons and those via RNA are referred to as retrotransposons.

- — *For transposons*: The transposon in donor DNA is cut by a special enzyme and then inserted into a target DNA. In bacteria, this enzyme is called transposase which has both nuclease and ligase activities.

— *For retrotransposons*: The retrotransposon in donor DNA is first transcribed into RNA and then reverse-transcribed into DNA, which is inserted into a target DNA by the same recombination mechanism as the

In bacteria, the target DNA is cut by transposase, producing sticky ends (single strands, easy to pair with complementary sequences). The transposase also has ligase activity which may ligate the intermediate DNA of transposons. The gaps at sticky ends are filled by DNA polymerase, generating direct repeats. Other organisms are expected to use the same mechanism to insert the DNA intermediates of either transposons or retrotransposons into target DNA. The direct repeats generated by this mechanism is found in ALL mobile elements. In recombinant DNA technology, a DNA fragment is inserted into a target DNA by a slightly different mechanism.

Holliday Model of DNA Crossover

The widely accepted model for genetic recombination was first proposed by Robin Holliday in 1964. The detailed mechanism of homologous recombination was mainly obtained from the study of E. coli. Although bacteria do no undergo meiosis, homologous recombination could occur during or immediately after DNA replication. It may also occur in a mating process called conjugation. In E. coli, the recombination is initiated by the enzyme RecBCD, consisting of three subunits: RecB, RecC and RecD. This enzyme has both helicase and nuclease activities.

The enzyme first uses its helicase activity to unwind DNA. When it hits the Chi site (with sequence GCTGGTGG), one of the exposed strand will be cut by its nuclease activity. The reason why this special site is called the "Chi site" is because the Greek letter χ (chi) looks like a crossover point. The Chi site is the position of the Holliday junction and also the position of chiasma. After DNA strands are cut by RecBCD, the strand invasion is catalysed by RecA proteins, which can wrap around single stranded DNA and direct it to form the Holliday structure. Finally, the branch migration is catalysed by RuvA and RuvB. The Holliday structure is resolved by the protein RuvC.

Gene Conversion

Gene conversion is a nonreciprocal transfer of genetic information.

Unequal Crossover

Due to significant sequence homology, the repetitive sequence region of a chromatid may not line up exactly with its corresponding region in a homologous chromatid or identical sister chromatid.

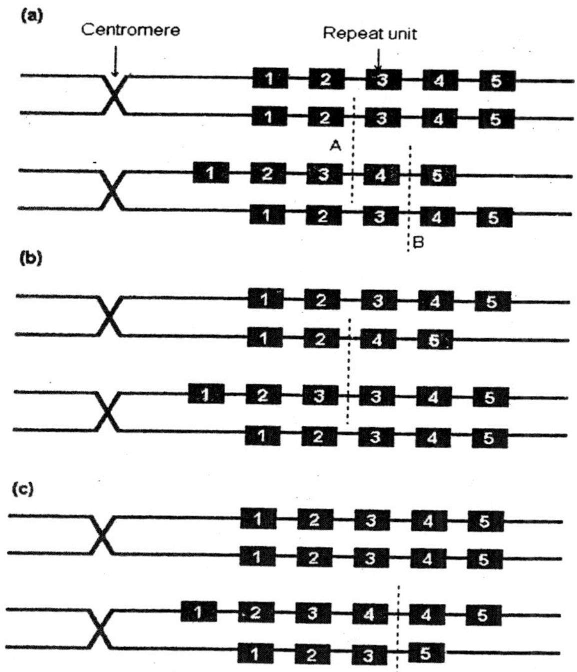

Figure 4. Unequal crossover and sister chromatid exchange.

As a result, different number of repeat units may be generated during meiosis. This is thought to be the major mechanism underlying VNTR.

a) Two pairs of sister chromatids line up during meiosis. A repetitive region of one chromatid (the third one) does not line up exactly with its corresponding region in other chromatids.

b) Strand breaks on nonsister chromatids (along line A) will result in unequal crossover, producing different number of repeat units in these chromatids.

c) Strand breaks on sister chromatids (along line B) also produce different repeats. In this case, it is called sister chromatid exchange. The detailed mechanism of DNA crossover in (b) and (c) may be explained by the Holliday model.

CHROMOSOME ABNORMALITY

The cell contains three haploid sets of chromosomes (69,XXX or 69,XXY) caused by one of the following mechanisms:

— Fertilisation of a single egg by two sperm.

— Fertilisation between a normal egg and an abnormal diploid sperm.

— Fertilisation between an abnormal diploid egg and a normal sperm.

Monosomy

The cell has only one copy of a specific chromosome. For example, individuals with Turner's syndrome have a single X chromosome and no Y.

Trisomy

The cell has an extra copy of a specific chromosome. Examples:

— Down's syndrome - with three copies of chromosome 21.

— Klinefelter's syndrome - with three sex chromosomes (XXY).

Both monosomy and trisomy arise from the same mechanism explained as:

a) During mitotic metaphase or meiotic metaphase II, one pair of sister chromatids does not align at the equatorial plane. Consequently, both sister chromatids move to the same cell. One daughter cell will show trisomy and another daughter cell will have monosomy.

b) During meiotic metaphase I, two homologous pairs of sister chromatids do not align at the equatorial plane and subsequently move to the same cell. The resulting daughter cells also have either monosomy or trisomy.

Structural Abnormality

Two breaks in a single chromosome can cause inversion, deletion or ring structure as shown in the following figure.

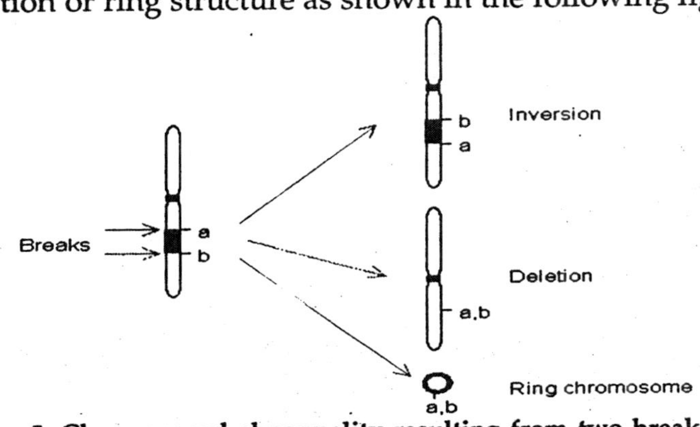

Figure 5. Chromosomal abnormality resulting from two breaks in a single chromosome.

Breaks in more than one chromosome can lead to translocation such as reciprocal translocation or Robertsonian translocation.

Inversion: The segment between two breakpoints is inverted before resealing the breaks.

Deletion: The breaks reseal without including the segment between breakpoints. Examples: cri-du-chat syndrome and William's syndrome.

Ring chromosome: Two ends of the segment between breakpoints are joined to form a circular structure.

DNA REPLICATION

DNA replication is the process of copying a double-stranded DNA molecule to form two double-stranded molecules. The process of DNA replication is a fundamental process used by all living organisms as it is the basis for biological inheritance. As each DNA strand holds the same genetic information, both strands can serve as templates for the reproduction of the opposite strand. The template strand is preserved in its entirety and the new strand is assembled from nucleotides — this process is called "semiconservative replication". The resulting double-stranded DNA molecules are identical; proofreading and error-checking mechanisms exist to ensure near perfect fidelity.

Each cell contains elaborate machinery for accurately copying its store of genetic information, as well as specialised enzymes for repairing DNA when it is damaged. These enzymes catalyse some of the most rapid and accurate processes that take place within cells, and their actions reflect the elegance and efficiency of cellular chemistry. Despite these systems for protecting the genetic instructions from copying errors and accidental damage, permanent changes, or mutations, sometimes do occur. Mutations in the DNA often affect the information it encodes.

Occasionally, this can benefit the organism in which a mutation occurs: for example, mutations can make bacteria resistant to antibiotics that are used to kill them. Indeed, the accumulation of changes in DNA over millions of years provides the variety in genetic material that makes one species distinct from another. Mutations also produce the smaller variations that underlie the differences between individuals of the same species that we can easily see in humans and other animals.

However, mutations are often detrimental: in humans, mutations are responsible for thousands of inherited diseases, and mutations that arise in the cells of the body throughout the lifetime of an individual may also cause disease, most notably the many types of cancer. Thus survival of a cell or organism can depend on preventing changes to its DNA. Without the cellular systems that are continually monitoring and repairing damage to DNA, it is questionable whether life could exist at all.

DNA replication and repair-that are responsible for keeping mutations to a minimum. Some of the intriguing ways in which cells alter their genetic information, including DNA recombination and the movement of the special DNA sequences in our chromosomes called transposable elements. At each cell division, a cell must copy its genome with extraordinary accuracy.

Enabling Base-Pairing

Each strand of the DNA double helix contains a sequence of nucleotides that is exactly complementary to the nucleotide sequence of its partner strand. Each strand can therefore act as a template, or mold, for the synthesis of a new complementary strand. In other words, if we designate the two DNA strands as S and S′, strand S can serve as a template for making a new strand S′, while strand S′ can serve as a template for making a new strand S. Thus, the genetic information in DNA can be accurately copied by the beautifully simple process in which strand S separates from strand S′, and each separated strand then serves as a template for the production of a new complementary partner strand that is identical to its former partner.

The ability of each strand of a DNA molecule to act as a template for producing a complementary strand enables a cell to copy, or replicate, its genes before passing them on to its descendants. But the task is aweinspiring, as it can involve copying billions of nucleotide pairs every time a cell divides. The copying must be carried out with speed and accuracy: in about 8 hours, a dividing animal cell will copy the equiva-lent of 1000 books like this one and, on average, get no more than a single letter or two wrong. This feat is performed by a cluster of proteins that together form a "replication machine."

DNA replication produces two complete double helices from the original DNA molecule, each new DNA helix identical (except for rare errors) in nucleotide sequence to the parental DNA double helix. Because each parental strand serves as the template for one new strand, each of the daughter DNA double helices ends up with one of the original (old) strands plus one strand that is completely new; this style of replication is said to be semiconservative.

DNA Synthesis

The DNA double helix is normally very stable: the two DNA strands are locked together firmly by the large numbers of hydrogen bonds between the bases on both strands. As a result, only temperatures approaching those of boiling water provide enough thermal energy to separate these strands. In order to be used as a template, however, the double helix must first be opened up and the two strands separated to expose unpaired bases. How does this occur at the temperatures found in living cells?

The process of DNA replication is begun by initiator proteins that bind to the DNA and pry the two strands apart, breaking the hydrogen bonds between the bases.

Although the hydrogen bonds collectively make the DNA helix very stable, individually each hydrogen bond is weak. Separating a short length of DNA does not therefore require a large energy input and can occur with the assistance of these proteins at normal temperatures.

The positions at which the DNA is first opened are called replication origins, and they are marked by a particular sequence of nucleotides. In simple cells like those of bacteria or yeast, replication origins span approximately 100 base pairs; they are composed of DNA sequences that attract the initiator proteins, as well as stretches of DNA that are especially easy to open. Therefore, DNA rich in A-T base pairs is relatively easy to pull apart, and A-T-rich stretches of DNA are typically found at replication origins. A bacterial genome, which is typically contained in a circular DNA molecule of several million nucleotide pairs, has a single origin of replication. The human genome, which is very much larger, has approximately 10,000 such origins.

In humans, beginning DNA replication at many places at once allows a cell to replicate its DNA relatively quickly. Once an initiator protein binds to DNA at the replication origin and locally opens up the double helix, it attracts a group of proteins that carry out DNA replication. This group operates as a protein machine, with each member carrying out a specific function. For eucaryotic cells, DNA replication is a monumental task. To transmit its genetic information to its daughters, a cell must copy its entire genome quickly, carefully, and completely. And it must copy its genetic material once and only once during each round of cell division.

Failure to perform these tasks properly-essentially perfectly-can have catastrophic consequences. If the whole genome is not replicated, critical information may be lost; copying some information more than once can be equally disastrous. Both situations can cause mutations, duplications, rearrangements, even massive chromosomal breakage-defects that in a multicellular organism can cause cancer or other diseases. But how do cells control this complicated molecular maneuver? Eucaryotic genomes contain many origins of replication-hundreds in simple yeasts, thousands in humans-and not all of these origins are equal. Some are used faithfully in each round of replication; others are used periodically, occasionally, or even rarely. Furthermore, different origins are called into service at different times during S phase, the stage in the cell cycle when DNA is replicated. Some "fire" early in S phase, others launch replication later. Their activity depends, at least in part, on where in the genome they are located.

With so many replication origins all over the genome, firing at different times, it is hard to imagine how a cell can coordinate the process with such extraordinary accuracy. Current studies aimed at identifying and locating replication origins, and determining when they are used, are beginning to reveal some answers as to how origins are chosen, how the timing of their activity is determined, and how the whole process is controlled.

Replication in Eucaryotes

Unlike the situation in procaryotes, origins of replication in eucaryotes do not contain an easily identifiable DNA sequence. Most replication origins in the yeast genome contain variations of an 11-nucleotide sequence, but not all DNA fragments with this sequence can act as origins. In other eucaryotes, almost any sequence seems capable of initiating replication. The simplest approach to determining whether a piece of DNA contains an origin is to see whether the fragment can direct the replication of a plasmid.

Plasmids are small loops of DNA that can exist inside a yeast or bacterial cell, separate from the cell's own genome. These circular DNA molecules have their own replication origins, which allow them to replicate independently of the host cell's genome. Using recombinant DNA techniques, an investigator can remove a plasmid's origin of replication and replace it with a DNA fragment that might contain a yeast replication origin. The modified plasmids are then introduced into a yeast cell and the cells are incubated to allow replication to occur: replication of the plasmid can only occur if its inserted DNA fragment contains an origin of replication. In this way, yeast origins of replication can be identified.

The plasmid maintenance method is fine when one starts with a piece of DNA-or a small collection of DNA fragments -that potentially contain a replication origin. But what if a researcher has no idea where the origins lie, or wishes to identify and study all of the origins in a eucaryote such as yeast? Now some powerful new techniques are allowing investigators to locate the replication origins across a whole genome-and determine when they fire-in one fell swoop.

To map the positions and timing of all the replication origins in the yeast genome, researchers have turned to DNA microarrays-grids studded with thousands of DNA fragments of known sequence. These microarrays are used to determine whether nucleotide fragments of a particular sequence are present in a sample. They are widely used, for example, to monitor which genes are being expressed in a cell. Because the exact sequence-and position-of every DNA fragment on the microarray is known, the exact sequence of any nucleotide fragment that binds to its complementary sequence in the array can be ascertained simply by determining to which position on the array it is bound.

For the replication origin experiment, researchers use a microarray that contains DNA fragments that cover the entire yeast genome-tens of thousands of probes that represent unique sequences found at consecutive intervals of, say, 500 nucleotides along the genome. The yeast are allowed to begin replication and their DNA is collected, broken into fragments, and labeled with a fluorescent marker. These fragments are washed over the whole-genome microarray and allowed to bind to their complementary sequences fixed to the surface of the array. Researchers can then locate the replication origins by seeing which DNA segments are duplicated first.

In reality, replication in yeast cells is initiated at hundreds of origins that fire at different times. Fortunately, the same experimental protocol can be adapted to determine which regions of DNA replicate first, which start later, and how quickly replication forks sweep through different segments of the genome. To follow replication over time, researchers collect DNA from yeast cells at different times after the start of replication. They begin by synchronising the yeast cells so that each cell in the population will begin to replicate its DNA at exactly the same time. This can be done by exposing the yeasts to a molecule that will cause them to arrest in G1, the phase of the cell cycle in which the replication machinery assembles on the origins but in which DNA synthesis has yet to begin. When the drug is removed, the yeast cells will all enter S phase.

DNA samples are collected, fluorescently labeled, and washed over the microarray as before. Now, however, researchers can monitor the progression of replication throughout the genome. Using such microarray-based techniques, researchers have identified some 300 origins of replication in the yeast genome; some of these had previously been identified using the plasmid maintenance methods described above, but many are new. They also found that replication origins are activated throughout S phase, with most firings occurring near mid-S phase. Although the experiment was performed with yeast DNA, this method can be adapted to look for replication origins in human DNA. Such studies should enhance our understanding of how all cells achieve such exquisite control over a process as critical and complex as DNA replication.

New DNA synthesis

DNA molecules in the process of being replicated can be observed in the electron microscope, where it is possible to see Yshaped junctions in the DNA, called replication forks. At these forks, the replication machine is moving along the DNA, opening up the two strands of the double helix and using each strand as a template to make a new daughter strand. Two replication forks are formed starting from each replication origin, and they move away from the origin in both directions, unzipping the DNA as they go. DNA replication in bacterial and eucaryotic chromosomes is therefore termed bidirectional. The forks move very rapidly-at about 1000 nucleotide pairs per second in bacteria and 100 nucleotide pairs per second in humans. The slower rate of fork movement in humans (indeed, in all eucaryotes) may be due to the difficulties in replicating through the more complex chromatin structure found in these higher organisms.

At the heart of the replication machine is an enzyme called DNA polymerase, which synthesises the new DNA using one of the old strands as a template. This enzyme catalyzes the addition of nucleotides to the 3′ end of a growing DNA strand by the formation of a phosphodiester bond between this end and the 5′-phosphate group of the incoming nucleotide. The nucleotides enter the reaction initially as energy-rich nucleoside triphosphates, which provide the energy for the polymerisation reaction. The hydrolysis of one phosphoanhydride bond in the nucleoside triphosphate provides the energy for the condensation reaction that links the nucleotide monomer to the chain and releases pyrophosphate (PPi).

The DNA polymerase couples the release of this energy to the polymerisation reaction. Pyrophosphate is further hydrolysed to inorganic phosphate (Pi), which makes the polymerisation reaction effectively irreversible. DNA polymerase does not dissociate from the DNA each time it adds a new nucleotide to the growing chain; rather, it stays associated with the DNA and moves along it stepwise for many cycles of the polymerisation reaction.

Asymmetrical Replication Fork

The 5′-to-3′ direction of the DNA polymerisation mechanism poses a problem at the replication fork. As a consequence, at the replication fork, one new DNA strand is being made on a template that runs in one direction (3′ to 5′), whereas the other new strand is being made on a template that runs in the opposite direction (5′ to 3′). The replication fork is therefore asymmetrical. Both the new DNA strands appear to be growing in the same direction, that is, the direction in which the replication fork is moving. On the face of it this suggests that one strand is being synthesised in the 3′-to-5′ direction and one is being synthesised in the 5′-to-3′ direction.

DNA polymerase, however, can catalyse the growth of the DNA chain in only one direction; it can add new subunits only to the 3′ end of the chain. As a result, a new DNA chain can be synthesised only in a 5′-to-3′ direction. This can easily account for the synthesis of one of the two strands of DNA at the replication fork, but not the other. One might have expected a second DNA polymerase to synthesise the other DNA strand-one that works by adding subunits to the 5′ end of a DNA chain.

However, no such enzyme exists. Instead, the problem is solved by the use of a "backstitching" maneuver. The DNA strand whose 5′ end must grow is made discontinuously, in successive separate small pieces, with the DNA polymerase working backward from the replication fork in the 5′-to-3′ direction for each new piece. These pieces-called Okazaki fragments after the biochemist who discovered them-are later "stitched" together to form a continuous new strand. The DNA strand that is synthesised discontinuously in this way is called the lagging strand; the strand that is synthesised continuously is called the leading strand.

Although they differ in subtle details, the replication forks of all cells, procaryotic and eucaryotic, have leading and lagging strands. The common feature arises from the fact that all of the DNA polymerases used to replicate DNA polymerise in the 5′-to-3′ direction only.

DNA Polymerase

DNA polymerase is so accurate that it makes only about one error in every 107 nucleotide pairs it copies. This error rate is much lower than can be accounted for simply by the accuracy of complementary basepairing. Although A-T and C-G are by far the most stable base pairs, other, less stable base pairs-for example, G-T and C-A-can also be formed. Such incorrect base pairs are formed much less frequently than correct ones, but they occur often enough that they would kill the cell through an accumulation of mistakes in the DNA if they were allowed to remain. This catastrophe is avoided because DNA polymerase can correct its mistakes. As well as catalysing the polymerisation reaction, DNA polymerase has an error-correcting activity called proofreading.

Before the enzyme adds a nucleotide to a growing DNA chain, it checks whether the previous nucleotide added is correctly base-paired to the template strand. If so, the polymerase adds the next nucleotide; if not, the polymerase removes the mispaired nucleotide by cutting the phosphodiester bond it has just made, releases the nucleotide, and tries again. Thus, DNA polymerase possesses both a 5′-to-3′ polymerisation activity and a 3′-

to-5′ exonuclease (nucleic acid-degrading) activity. These activities are carried out by different domains within the polymerase molecule.

This proofreading mechanism explains why DNA polymerases synthesise DNA only in the 5′-to-3′ direction, despite the need this imposes for a cumbersome backstitching mechanism at the replication fork. A hypothetical DNA polymerase that synthesised in the 3′-to-5′ direction (and would thereby circumvent the need for backstitching) would be unable to proofread: if it removed an incorrectly paired nucleotide, the polymerase would create a chain end that is chemically dead, in the sense that it would no longer be able to elongate. Thus, for a DNA polymerase to function as a self-correcting enzyme that removes its own polymerisation errors as it moves along the DNA, it must proceed only in the 5′-to-3′ direction.

Short Lengths of RNA

We have seen that the accuracy of DNA replication depends on the requirement of the DNA polymerase for a correctly base-paired end before it can add more nucleotides. But since the polymerase can join a nucleotide only to a base-paired nucleotide in a DNA double helix, it cannot start a completely new DNA strand. A different enzyme is needed to begin a new DNA strand, an enzyme that can begin a new polynucleotide chain simply by joining two nucleotides together without the need for a base-paired end. This enzyme does not, however, synthesise DNA. It makes a short length of a closely related type of nucleic acid-RNA (ribonucleic acid)-using the DNA strand as a template.

This short length of RNA, around 10 nucleotides long, is base-paired to the template strand and provides a base-paired 3′ end as a starting point for DNA polymerase. It thus serves as a primer for DNA synthesis, and the enzyme that synthesises the RNA primer is known as primase. A strand of RNA is very similar chemically to a single strand of DNA except that it is made of ribonucleotide subunits, in which the sugar is ribose, not deoxyribose; RNA also differs from DNA in that it contains the base uracil (U) instead of thymine (T). However, since U can form a base pair with A, the RNA primer is synthesised on the DNA strand by complementary base-pairing in exactly the same way as is DNA.

For the leading strand, an RNA primer is needed only to start replication at a replication origin; once a replication fork has been established, the DNA polymerase is continuously presented with a basepaired 3′ end as it tracks along the template strand. As the movement of the replication fork exposes a new stretch of unpaired bases, a new RNA primer is made at intervals along the lagging strand. DNA polymerase adds a deoxyribonucleotide to the 3′ end of this primer to start a DNA strand, and it will continue to elongate this strand until it runs into the next RNA primer.

To produce a continuous new DNA strand from the many separate pieces of RNA and DNA made on the lagging strand, three additional enzymes are needed. These act quickly to remove the RNA primer, replace it with DNA, and join the DNA fragments together: a nuclease breaks apart the RNA primer, a DNA polymerase called a repair polymerase replaces the RNA with DNA (using the adjacent Okazaki frag-ment as a primer), and the enzyme DNA ligase joins the 5′-phosphate end of one new DNA fragment to the 3′-hydroxyl end of the next. ATP or NADH is required for ligase activity.

Primase can begin new polynucleotide chains, but this is possible because it does not proofread its work. As a result, primers contain a high frequency of mistakes. But since they are made of RNA instead of DNA, the primers stand out as "suspect copy" to be automatically removed and replaced by DNA. This DNA is put in by DNA repair polymerases, which, like the replicative polymerases, proofread as they synthesise. In this way, the cell's replication machinery is able to begin new DNA chains and, at the same time, ensure that all of the DNA is copied faithfully.

Proteins at a Replication Fork

DNA replication requires a variety of proteins that act in concert with DNA polymerase. The additional proteins that, together with DNA polymerase and primase, form the protein machine that powers the replication fork forward and synthesises new DNA behind it. Although it would make good sense for the three proteins that replace RNA primers with DNA-nuclease, repair polymerase, and ligase-to also be a part of the replication machine, it is not yet known whether this is the case.

At the head of the replication machine is a helicase, a protein that uses the energy of ATP hydrolysis to speed along DNA, opening the double helix as it moves. Another component of the replication machine-single-strand binding protein-clings to the single-stranded DNA exposed by the helicase and transiently prevents it from re-forming base pairs. Yet another protein, called a sliding clamp, keeps the DNA polymerase firmly attached to the DNA template; on the lagging strand, the sliding clamp releases the polymerase from the DNA each time an Okazaki fragment is completed. This clamp protein forms a ring around the DNA helix and binds polymerase, allowing it to slide along a template strand as it synthesises new DNA.

Most of the proteins involved in DNA replication are thought to be held together in a large multienzyme complex that moves as a unit along the DNA, enabling DNA to be synthesized on both strands in a coordinated manner. This complex can be likened to a tiny sewing machine composed of protein parts and powered by nucleoside triphosphate hydrolysis. Although the structures of the individual protein components of the replication machine have been determined, how these components fit together is not known in detail. Some ideas about the general appearance of the complex, however, have been proposed.

Telomerase Replication

DNA replication begins at origins and how movement of the replication fork proceeds, we now turn to the special problem of replicating the very ends of eucaryotic chromosomes. The fact that DNA is synthesized only in the 5′-to-3′ direction means that the lagging strand of the replication fork is synthesized in the form of discontinuous DNA fragments, each of which is primed with an RNA primer laid down by a separate enzyme. When the replication fork approaches the end of a chromosome, however, the replication machinery encounters a serious problem: there is no place to lay down the RNA primer needed to start the Okazaki fragment at the very tip of the linear DNA molecule. Therefore, some DNA could easily be lost from the ends of a DNA molecule each time it is replicated.

Bacteria solve this "end-replication" problem by having circular DNA molecules as chromosomes. Eucaryotes solve it by having special nucleotide sequences at the ends of their chromosomes which are incorporated into telomeres. These repetitive telomeric DNA sequences attract an enzyme called telomerase to the chromosome. Telomerase adds multiple copies of the same telomere DNA sequence to the ends of the chromosomes, thereby producing a template that allows replication of the lagging strand to be completed. In addition to allowing replication of chromosome ends, telomeres serve additional functions: for example, the repeated telomere DNA sequences, together with the regions adjoining them, form structures that are recognized by the cell as the true ends of chromosomes rather than breaks that sometimes occur in the middle of chromosomes and must be repaired.

DNA REPAIR

The diversity of living organisms and their success in colonizing almost every part of the Earth's surface depends on genetic changes accumulated gradually over millions of years, allowing organisms to adapt to changing conditions and to colonize new habitats. However, in the short term, and from the perspective of an individual organism, genetic change is often detrimental, especially in multicellular organisms, where a genetic change can upset an organism's extremely complex and finely tuned development and physiology. To survive and reproduce, individuals must be genetically stable. This stability is achieved not only through the extremely accurate mechanism for replicating DNA, but also through mechanisms for correcting the rare copying mistakes made by the replication machinery and for repairing the accidental damage that is continually occurring to the DNA. Most of these changes in DNA are only temporary because they are immediately corrected by processes collectively called DNA repair.

Mutations

Cell's DNA replication and repair processes fail and allow a permanent change in the DNA. Such a permanent change is called a mutation, and it can have profound consequences. A mutation affecting just a single nucleotide pair can severely compromise an organism's fitness if the change occurs in a vital position in the DNA sequence. Because the structure and activity of a protein depend on its amino acid sequence, a protein with an altered sequence may function poorly or not at all.

For example, humans use the protein hemoglobin to transport oxygen in the blood; the sequence of nucleotides that encodes the amino acid sequence of one of the two types of protein chains (the β-globin chain) of the hemoglobin molecule. A permanent change in a single nucleotide in this sequence can cause cells to make a b-globin chain with an incorrect sequence of amino acids. Such a mutation causes the disease sickle-cell anemia. The sickle-cell hemoglobin is less soluble than normal hemoglobin and forms fibrous precipitates, which lead to the characteristic sickle shape of affected red blood cells. Because these cells are more fragile and frequently break in the bloodstream, patients with this potentially life-threatening disease have a reduced number of red blood cells, a deficiency that can cause weakness, dizziness, headaches, pain, and total organ failure.

The example of sickle-cell anemia, which is an inherited disease, illustrates the importance of protecting reproductive cells (germ cells) against mutation. A mutation in one of these will be passed on to all the cells in the body of the multicellular organism that develops from it, including the germ cells for production of the next generation. However, the many other cells in a multicellular organism (its somatic cells) must also be protected from genetic change to safeguard the health and well-being of the individual.

Nucleotide changes that occur in somatic cells can give rise to variant cells, some of which grow in an uncontrolled fashion at the expense of the other cells in the organism. In the extreme case, an uncontrolled cell proliferation known as cancer results. This disease, which is responsible for about 30% of the deaths that occur in Europe and North America, is due largely to a gradual accumulation of changes in the DNA sequences of somatic cells that is caused by random mutation.

Increasing the mutation frequency even two- or threefold would cause a disastrous increase in the incidence of cancer by accelerating the rate at which somatic cell variants arise. Thus the high fidelity with which DNA sequences are replicated and maintained is important both for the reproductive cells, which transmit the genes to the next generation, and for the somatic cells, which normally function as carefully regulated members of the complex community of cells in a multicellular organism.

DNA Mismatch Repair System

The cell has a backup system-called DNA mismatch repair-which is dedicated to correcting these rare mistakes. The replication machine itself makes approximately one error per 107 nucleotides copied; DNA mismatch repair corrects 99% of these errors, increasing the overall accuracy to one mistake in 109 nucleotides copied. This level of accuracy is much higher than that generally encountered in the visible world around us.

Whenever the replication machinery makes a copying mistake, it leaves a mispaired nucleotide (commonly called a mismatch) behind. If left uncorrected, the mismatch will result in a permanent mutation in the next round of DNA replication. A complex of mismatch repair proteins recognizes these DNA mismatches, removes (excises) one of the two strands of DNA involved in the mismatch, and resynthesizes the missing strand. To be effective in correcting replication mistakes, this mismatch repair system must always excise only the newly synthesized DNA strand: excising the other strand (the old strand) would preserve the mistake instead of correcting it.

In eucaryotes, it is not yet known for certain how the mismatch repair machinery distinguishes the two DNA strands. However, there is evidence that newly replicated DNA strands-both leading and lagging -are preferentially nicked; it is these nicks (single-stranded breaks) that appear to provide the signal that directs the mismatch repair machinery to the appropriate strand. The importance of mismatch repair in humans was recognized recently when it was discovered that an inherited predisposition to certain cancers (especially some types of colon cancer) is caused by a mutation in the gene responsible for producing one of the mismatch repair proteins.

Humans inherit two copies of this gene (one from each parent), and individuals who inherit one damaged mismatch repair gene show no symptoms until the undamaged copy of the gene is accidentally mutated in a somatic cell. This gives rise to a clone of somatic cells that, because they are deficient in mismatch repair, accumulate mutations more rapidly than do normal cells. Because most cancers arise from cells that have accumulated multiple mutations, a cell deficient in mismatch repair has a greatly enhanced chance of becoming cancerous. Thus, inheriting a damaged mismatch repair gene predisposes an individual to cancer.

Rare mistakes in DNA replication, can be corrected by the mismatch repair mechanism. There are also other ways in which the DNA can be damaged, and these require other mechanisms for their repair. Just like any other molecule in the cell, DNA is continually undergoing thermal collisions with other molecules. These often result in major chemical changes in the DNA. For example, during the time it takes to read this sentence, a total of about a trillion (10^{12}) purine bases (A and G) will be lost from the DNA of your cells by a spontaneous reaction called depurination.

Depurination does not break the phosphodiester backbone but, instead, gives rise to lesions that resemble missing teeth. Another major change is the spontaneous loss of an amino group (deamination) from cytosine in DNA to produce the base uracil. Some chemically reactive by-products of metabolism also occasionally react with the bases in DNA, altering them in such a way that their base-pairing properties are changed.

These are only a few of many chemical changes that can occur in our DNA. If left unrepaired, many of them would lead either to the substitution of one nucleotide pair by another as a result of incorrect basepairing during replication or to deletion of one or more nucleotide pairs in the daughter DNA strand after DNA replication. Some types of DNA damage (thymine dimers, for example) often stall the DNA replication machinery at the site of the damage. All of these types of damage, if unrepaired, would have disastrous consequences for an organism.

Stability of Genes

The thousands of random chemical changes that occur every day in the DNA of a human cell, by metabolic accidents or exposure to DNA-damaging chemicals, are repaired by a variety of mechanisms, each catalyzed by a different set of enzymes. Nearly all these mechanisms depend on the existence of two copies of the genetic information, one in each strand of the DNA double helix: if the sequence in one strand is accidentally damaged, information is not lost irretrievably, because a backup version of the altered strand remains in the complementary sequence of nucleotides in the other strand.

Most damage creates structures that are never encountered in an undamaged DNA strand; thus the good strand is easily distinguished from the bad.

As indicated, it involves three steps:

1. The damaged DNA is recognized and removed by one of a variety of different nucleases, which cleave the covalent bonds that join the damaged nucleotides to the rest of the DNA molecule, leaving a small gap on one strand of the DNA double helix in this region.
2. A repair DNA polymerase binds to the 3′-hydroxyl end of the cut DNA strand. It then fills in the gap by making a complementary copy of the information stored in the undamaged strand. Although a different enzyme from the DNA polymerase that replicates DNA, a repair

DNA polymerase synthesizes DNA strands in the same way. For example, it synthesizes chains in the 5′-to-3′ direction and has the same type of proofreading activity to ensure that the template strand is accurately copied. In many cells, this is the same enzyme that fills in the gap left after the RNA primers are removed in normal DNA replication.

3. When the repair DNA polymerase has filled in the gap, a break remains in the sugar-phosphate backbone of the repaired strand. This nick in the helix is sealed by DNA ligase, the same enzyme that joins the lagging-strand DNA fragments during DNA replication. Steps 2 and 3 are nearly the same for most types of DNA repair, including mismatch repair. However, step 1 uses a series of different enzymes, each specialized for removing different types of DNA damage. The importance of these repair processes is indicated by the large investment that cells make in DNA repair enzymes. Single-celled organisms such as yeasts contain more than 50 different proteins that function in DNA repair, and DNA repair pathways are likely to be even more complex in humans.

The importance of these DNA repair processes is also evident from the consequences of their malfunction. Humans with the genetic disease xeroderma pigmentosum, for example, cannot repair thymine dimers because they have inherited a defective gene for one of the proteins involved in this repair process. Such individuals develop severe skin lesions, including skin cancer, because of the accumulation of thymine dimers in cells that are exposed to sunlight and the consequent mutations that arise in the cells that contain them.

High Fidelity of DNA Maintenance

DNA is replicated and maintained with remarkable fidelity. As a consequence, changes in the DNA accumulate remarkably slowly in the course of evolution. Of course, the rate of evolutionary change in the DNA of a species depends also on the effects of natural selection: DNA copying errors that have harmful consequences for the organism are eliminated from the population through the death or reduced fertility of individuals carrying the misreplicated DNA. But the mechanisms of DNA replication and repair are so accurate that even where no such selection operates-at the many sites in the DNA where a change of nucleotide has no effect on the fitness of the organism-the genetic message is faithfully preserved over tens of millions of years.

Thus humans and chimpanzees, after about 5 million years of divergent evolution, still have DNA sequences that are at least 98% identical. Even humans and whales, after 10 or 20 times this period, still have chromosomes that are unmistakably similar in their DNA sequence and many proteins with amino acid sequences that are almost identical.

REFERENCES

Alberts B, Johnson A, Lewis J, Raff M, Roberts K, Walter P., "Chapter 5: DNA Replication, Repair, and Recombination", *Molecular Biology of the Cell*, Garland Science, 2002.

Berg J.M, Tymoczko J.L, Stryer L, Clarke N.D., "Chapter 27, Section 4: DNA Replication of Both Strands Proceeds Rapidly from Specific Start Sites", *Biochemistry*, W.H. Freeman and Company, 2002.

Lodish H, Berk A, Matsudaira P, Kaiser CA, Krieger M, Scott MP, Zipursky SL, Darnell J., *Molecular Cell Biology*, WH Freeman: New York, NY. 5th ed., 2004.

Morgan D.O., *The Cell Cycle: Principles of Control*, New Science Press: London, 2007.

3

Basics of Cell Functioning

Each cell needs to obtain oxygen and other nutrients (carbohydrates, amino acids, lipid molecules, minerals, etc.) from the environment, maintain water balance with its surroundings, and remove waste materials from the cell. The boundary between any cell and its environment (through which substances must pass) is the plasma membrane, composed of phospholipid and protein molecules.

PLASMA MEMBRANE

The plasma membrane has a number of functions for a cell.

— It serves as the boundary between the cytoplasm of the cell and the external environment, and selectively isolates the cell from the external environment.

— Maintains the cell's environment by regulating materials that enter or leave the cell.

— Provides mechanisms for cell-to-cell communication.

— Genetically unique cell recognition markers embedded in the plasma membrane provide mechanisms for a cell to recognize itself and other cells of its particular individual organism versus non-self.

This is important to the immune system and defense of the organism. Although the plasma membrane forms the boundary of the cell, and surrounds the cell, many internal structures of most cells also have their own membrane boundaries.

Fluid Mosaic Membrane Structure

The typical membrane structure consists of a phospholipid bilayer with a number of proteins scattered throughout, along with some carbohydrates (glycoproteins), glycolipids and sterols, similar to the way in which one does a mosaic tile, hence the name.

Phospholipid

A phospholipid has both polar and non-polar regions. The fatty acid "tails" of the two phospholipid layers are oriented towards each other so that the hydrophilic "heads", which contain the phosphate portion, face out to the environment as well as into the cytoplasm of the cell's interior, where they form hydrogen bonds with surrounding water molecules. Because the individual phospholipid molecules are not bonded to each other, a membrane is flexible (or "fluid"), something which is pretty important to its functions.

The fluidity of a membrane is crucial to its function. In caribou, circulation is reduced in the lower legs to prevent excess heat loss during cold winters. The membranes of the lower legs have more unsaturated fatty acids than those of the upper legs to retain more fluidity in reduced temperatures. Brain cell membranes in ground squirrels become more solid during hibernation. Phospholipids containing more polyunsaturated fatty acids are more fluid than those with fewer polyunsaturated fatty acids.

Cholesterol in membranes reduces fluidity as well. There are times when membranes need more or less fluidity, and molecular composition provides for sure membrane flexibility. Many materials that enter or leave the cell are water-soluble; the fatty acid layers serve as a barrier to their free entry. Proteins in the membrane are required to move these substances through the membrane. Lipids generally pass through the membrane more easily.

Membrane Proteins

The membrane proteins have a number of functions. Some are embedded in the phospholipid layers; others move (literally) throughout the membrane layers. Other membrane proteins are complexed to carbohydrate molecules, forming glycoproteins. Generally, there are a number of different membrane proteins.

Cell Adhesion Proteins

Some proteins are responsible for the cell junctions such as tight junctions that permit cells to adhere to each other. Some adhesion proteins attach to the cytoskeleton or extracellular matrix to help maintain cell shape (particularly for animal cells) and fix into position some membrane proteins. Collagen is an important glycoprotein of the extracellular matrix.

Communication Proteins

Communication proteins are responsible for the cell gap junctions that permit cells to communicate with each other. Gap junctions are common in heart muscle cells so that the contraction can be coordinated.

Receptor Proteins

Receptor proteins serve as binding or attachment sites, especially for hormones or other molecular messengers. Once activated, they trigger certain cell responses, often using signal transduction pathways. Some receptors work in conjunction with carrier proteins, opening gated channels for ion movement.

Other receptors are sensitive to nutrient levels, affect metabolic rate or even affect cell division.

Transport Proteins

Transport Proteins function as carriers, which have binding sites that attract specific molecules. When a molecule binds to the carrier protein, the protein changes shape and moves the substance through the membrane. This process often requires energy (ATP), and the ATP complex is a part of the transport protein. When ATP is involved with actively moving molecules through the protein channel the process is called Active Transport. Most of our ions (Ca^{++}, Na^{+}, Cl^{-}, K^{+}, etc.), along with amino acids, sugars and other small nutrient molecules are moved through transport proteins.

Other transport proteins form channels within the phospholipid bilayer, which allows small water-soluble molecules to pass through. Aquaporins are important water channels that facilitate the movement of water through membranes. Some channels are gated.

Recognition Proteins

Glycoproteins (carbohydrate-protein hybrids) and some glycolipids serve as surface receptors for cell recognition and identification. They are important to the immune system so that immune system cells can distinguish between one's own cells and foreign cells. Recognition proteins are also used to guide cell attachments/adhesions in developmental processes.

Enzymatic Proteins

Many enzymes are embedded in membranes, which attract reacting molecules to the membrane surface. Enzymes needed for metabolic pathways can be aligned adjacent to each other to act like an assembly line for the reactions.

DIFFUSION

Diffusion is the net movement of a substance from where there is more of it along a concentration gradient to where there is less of it, until molecules are equally distributed (and the gradient no longer exists).

Gradients are important in moving materials through membranes, both passively (without the use of energy by the cell) and actively (transport requiring cell energy).

Passive Transport involves moving things through membranes without the expenditure of cell energy down gradients. Passive transport in cells involves the process of diffusion.

Simple Diffusion

- Diffusion is a means of passive transport, since no additional energy is expended for the process.
- In terms of cellular activity, diffusion:
  - Requires no energy
  - But the cell has no control over diffusion, and the rate of diffusion is pretty slow and can not cover much distance.
  - The Rate of Diffusion can be affected by:
- Temperature (Higher temperature, faster molecule movement)
- Molecule size (Smaller molecules often move more easily)
- Concentration (Initial rate faster with higher concentration)
- Electrical and pressure gradients of the two regions (Greater the gradient differential, the more rapid the diffusion (again, initially)) Materials that may move through membranes by simple diffusion include:
  - H_2O (water)

  - CO_2 (carbon dioxide)
  - O_2 (oxygen)
- Some small lipid-soluble molecules (alcohol)

Facilitated Diffusion

Most molecules can not move freely through the membrane, but do cross membranes with the help of membrane transport proteins, which temporarily bind to the substance to be moved through the membrane, a process called facilitated diffusion or passive transport. No energy is involved, so it is still a passive process. Both carrier proteins and channel proteins are involved in facilitated diffusion.

Materials that move through membranes by facilitated diffusion include:

- Glucose
- Many small ions
- Amino acids

The movement of water through membranes also involves facilitated diffusion. There are special channel proteins, called aquaporins that facilitate the movement of water at a rate needed for cell activities.

TRANSPORTING MATERIALS

A significant part of membrane activity involves transporting materials through it in one direction or the other. Recall that the plasma membrane is selectively, or differentially, permeable. This means that:

- Some materials freely pass—the membrane is permeable to such molecules and whether they are inside or outside of the cell depends on other factors
- Some materials are excluded
- Some materials enter or leave the cell only by the using cell energy

For example, small hydrophobic molecules, such as CO_2, O_2 and small lipids, dissolve in the membrane and pass through readily. Tiny polar molecules, such as H_2O and alcohol, can also slip between the phospholipid molecules. Ions and most nutrient molecules do not move freely through membranes, but are often carried by the transport protein channels, either with or without the use of energy. Most large molecules are excluded and must be manufactured within the cell, or moved by significant alterations of the membrane itself.

All cells need to move some substances through membrane in a direction counter to the gradient, or move substances that are too large or bulky be moved without the use of cell energy. Cells have a number of ways to move things with the use of energy. Some transport proteins (carrier proteins) can move substances through the membrane against the concentration gradient. Active transport typically requires two carrier protein active sites: one to recognize the substance to be carried, and one to release ATP to provide the energy for the protein carriers or "pumps". Much energy is expended by the cell to do this!

In other cases, concentration gradients of ions, typically H^+ or Na^+ ions, can be used to provide the energy needed to move something through a membrane. For example, the substance to be moved is "coupled" to the concentration of H^+, and while to H^+ is moving "down" through the carrier channel, the substance is transported "up".

Larger substances may require changes in membrane shape and the fusion of the plasma membrane with vesicles containing the substances to successfully move the needed substance. Such changes in membranes occur throughout the lifetime of the cell, so that membrane sections are constantly being formed and reformed in the cell. Movement out of the cell involving changes of the membranes and formation of vesicles is exocytosis. Movement of materials into the cell is endocytosis.

ENDOCYTOSIS

Substances that enter the cell in this manner move by endocytosis. There are a variety of endocytosis processes: pinocytosis, receptor mediated endocytosis and phagocytosis.

Methods of Endocytosis

Pinocytosis

Membrane invaginates, substances "fall" in cavity, used for moving fluids into or out of a cell.

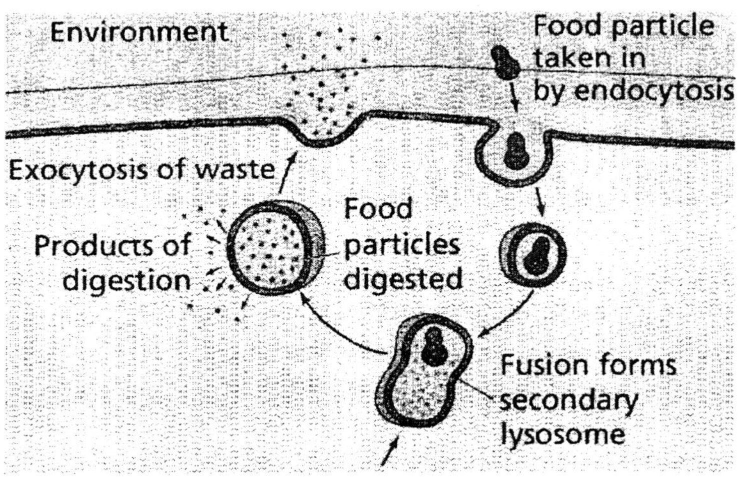

Endocytosis and exocytosis

Whatever molecules were in the fluid will be moved into the cell.

Receptor-Mediated Endocytosis

Receptor molecules in the membrane attract the substance to be moved into the cell, creating a membrane depression in that area (or coated pit). When sufficient molecules have been attracted, the pocket will be pinched off forming a coated vesicle in the cytoplasm.

Phagocytosis

Membrane surrounds and engulfs object, used for larger objects, such as engulfing prey by the protist, Amoeba, and bacteria by white blood cells.

Osmosis

Osmosis is the movement (diffusion) of water across a differentially permeable membrane in response to solute (dissolved substances) gradients that are maintained by the membrane. The "force" to move water through membranes is called osmotic pressure. It is comparable to physical pressure. Osmotic pressure may be resisted by the cell membrane (if it is strong enough) or the cell wall, in organisms that have cell walls. The wall or membrane exerts a mechanical pressure. The difference in the osmotic pressure and the wall or membrane pressure is known as water potential. Water potential is very important in a number of processes.

For the process of osmosis:

— The membrane is permeable to water.

— The membrane is not permeable to the solute(s), and the solutes will be substances which can "bind" to water, affecting the free flow of water.

A water gradient exists, in part because dissolved substances always lower the concentration of water in a solution. Since osmosis depends of the differences in the concentration of water, the specific types of solutes do not matter; it's their collective effect on the concentration of water than counts.

Hypertonic The solution has a higher solute concentration (less water) than the cell Hypertonic solutions will cause water to leave cells by osmosis, and cells may shrink.

Hypotonic The solution has a lower solute concentration (more water) than the cell Hypotonic solutions will cause water to enter cells by osmosis, causing the cells to swell.

Isotonic Equal proportions of solutes to water on both sides of the membrane.

Although we do not need to remember the specific terms which describe the ratio of solutes to water inside and outside of cells, we do need to understand the impact of these different conditions on how cells function. Cells can not afford to either lose water, or gain excess water. They must maintain an equal proportion of solutes both inside and outside of the cells, a condition called osmotic balance, to function. The process by which organisms regulate their osmotic balance is called osmoregulation.

PROTEASE ACTIVATED RECEPTORS

Cell surface receptors are activated by ligand binding. This interaction induces a conformational change in the receptor that results in a signal being transmitted across the cell membrane. The activation is reversible in that the ligand can dissociate from its receptor. Seven transmembrane domain G-protein coupled receptors (GPCR) are perhaps the most important and well-characterized type of receptor, providing the basis for numerous pharmacological interventions.

In recent years, it has become clear that GPCRs mediate the actions of almost all types of chemical messengers from small transmitters (catecholamines, histamine, acetylcholine, etc.), lipid messengers (prostaglandins, leukotrienes, platelet activating factor, etc.), neuropeptides (opiates, neuropeptide Y, tachykinins, etc.), peptide hormones (glucagon, angiotensin, calcitonin, etc.), to large glycoproteins (thyroid-stimulating hormone, luteinizing hormone, etc.).

Within this family of GPCRs, a different and unique mechanism of activation was discovered with the cloning of the thrombin receptor. This novel receptor was termed protease-activated receptor-1 (PAR-1). PAR-1 is activated by proteolytic cleavage at a specific site within the extracellular amino-terminal domain, exposing a new amino terminus, which is then thought to interact with a distal region of the receptor, leading to receptor activation. The discovery of this receptor has provided a framework for understanding how thrombin communicates with cells and provides a new pharmacological target for controlling thrombin's cellular actions. This unusual receptor exemplifies an emerging family of GPCRs that are activated by proteolysis. Given that proteases are very common enzymes in organisms, the catalytic activation of these receptors would seem to be an efficient means to regulate cellular responses.

Indeed, many protease-activation mechanisms have been identified in biology, suggesting an important role for protease activated receptors. PAR-1. PAR-1 exhibits an extended amino terminus containing both a putative thrombin cleavage site $LDPR^{41}/S^{42}$) and a hirudin-like domain (K^{52}YEPF) complementary to thrombin's anion binding exosite. Thrombin binds to PAR-1 with increased affinity through this latter site and subsequently cleaves the extended amino terminus.

This mechanism of receptor activation is supported by studies utilizing synthetic peptides representing this new amino terminus: SFLLRN for human PAR-1, SFFLRN for hamster and rat PAR-1. Agonist peptides can directly activate recombinant as well as native receptor. Additional structural features of the receptor have been identified from receptor modeling and mutagenesis studies. The second extracellular loop contains several residues important for ligand/receptor interaction and thus represents an important site for targeted receptor modulators. The intracellular loops are recognized sites for G-protein coupling and the cytoplasmic tail contains numerous sites for receptor phosphorylation leading to receptor inactivation and internalization.

PAR-1 from both hamster and rat were also cloned and shown to be highly homologous to human PAR-1. PAR-1 is widely distributed among cells and tissues, consistent with thrombin's widespread effects described in in vitro studies. Thus, regulation of PAR-1 expression and/or activity presents an important approach to modulating thrombin's actions. PAR-3 and PAR-4. Since the discovery of PAR-1, three additional PARs have been discovered: PAR-2, PAR-3, and PAR-4.

PAR-2 was first identified from a mouse genomic library using hybridization probes derived from the bovine substance K receptor. The human gene was subsequently isolated from a human genomic library and cloned from human kidney cDNA. PAR-2 is the only PAR described to date that is not activated by thrombin. The most widely studied proteolytic activators of PAR-2 are trypsin and tryptase, however several other serine proteases have recently been shown to activate PAR-2 including membrane-type serine protease-1, factor VIIa, and factor Xa. The cell and tissue distribution of PAR-2, like PAR-1, is widespread with significant expression in pancreas, colon, kidney, small intestine, and liver. PAR-2 is notably absent from platelets.

PAR-3 was first described in PAR-1 deficient mice whose platelets were fully responsive to thrombin. PAR-3 contains a consensus sequence for the thrombin anion-binding site indicating that it is a putative substrate for thrombin cleavage. However, PAR-3 is unique among the PARs in that it is not activated by synthetic peptides representing its tethered peptide ligand. Thus the functional role of PAR-3, other than in murine platelet activation, remains unclear. Human and murine PAR-3 display distinct tissue distributions based on Northern analysis. Human PAR-3 is expressed in several tissues including heart, liver, pancreas, small intestine, stomach, bone marrow, and vascular endothelial cells but is absent from human platelets. In contrast, megakaryocytes in the spleen were the only primary site of PAR-3 expression in mouse tissues.

Human PAR-4 was discovered using a bioinformatics approach employing est sequences homologous to other PARs. Concurrently, murine PAR-4 was described in studies with PAR-3 deficient mice. These mice, while devoid of thrombin sensitive responses at low nanomolar thrombin concentrations, were still responsive to higher thrombin concentrations. These observations led to a search for an additional thrombin receptor, culminating in the identification of murine PAR-4. PAR-4, unlike both PAR-1 and PAR-3, does not possess a consensus sequence for binding of thrombin's anionic exosite. Human PAR-4 is activated by approximately 100-fold higher thrombin concentrations than PAR-1. Part of this discrepancy may be attributable to the lack of this binding site on the receptor, which is important for designating thrombin's specificity and potency. The separation in thrombin potency between PAR-1 and PAR-4 implies that they may serve distinct roles in mediating cell function. Alternatively, for human PAR-4 other proteases such as trypsin or cathepsin G may be the primary activators.

PARS IN THROMBOSIS

Platelets are the principal cell type participating in the regulation of hemostasis. They are activated by a variety of stimuli including collagen, ADP, thromboxane A2, epinephrine, serotonin, and thrombin, but thrombin is considered the most potent stimulus of platelets. Thrombin induces platelet shape change, degranulation, aggregation, and biosynthesis of lipids, all of which contribute to the generation of thrombi. While these events are important during normal hemostasis to maintain the integrity of the vasculature, increased generation of thrombin and formation of large thrombi that lead to vessel occlusion contribute to the often fatal outcome of myocardial infarction, angina, and stroke.

Thus, understanding the mechanism by which thrombin can activate platelets may provide important therapeutic approaches to treating thrombotic related diseases. The presence of both PAR-3 and PAR-4 in murine platelets led Kahn et al. to propose a dual thrombin receptor activating system in platelets. The thrombin sensitivity in murine platelets in wild type versus PAR-3 deficient mice raised the intriguing possibility that PAR-3 and PAR-4 interacted in some way to signal as PAR-1 does in human platelets. Indeed, recent studies described the cooperativity of the two receptors whereby PAR-3 acted in concert with PAR-4 by localizing thrombin to the cell surface, allowing PAR-4 to be cleaved and activated.

The putative role of the PAR-3 tethered peptide ligand remains elusive however. The functionality of PAR-3 alone in murine platelets awaits the generation of a PAR-4 deficient mouse. A dual mechanism of activation has also been proposed for PAR-1 and PAR-4 in human platelets. Interestingly, PAR-4 activation in human platelets has been shown, however PAR-3 has not been identified in human platelets via sensitive methods such as reverse transcriptase-polymerase chain reaction. These observations suggest that PAR pairing seems to be the preferred system by which platelets respond to thrombin, PAR-1/PAR-4 in human platelets and PAR-3/PAR4 in murine platelets.

In other tissues however, the functional role of PAR-3 and PAR-4 remains to be fully characterized. PAR-3 has been described in endothelial cells for example, but PAR-4 has not. Through the use of PAR specific probes and antibodies, the tissue expression patterns for these receptors will be described, ultimately leading to studies that better define the functional role of these additional thrombin responsive receptors. Clearly, PAR-1 mediates the majority of thrombin's actions on platelets due to its sensitivity to low concentrations of thrombin.

Inhibition of PAR-1 in isolated human platelets was obtained with receptor blocking antibodies or the PAR-1 antagonist BMS-200261. These agents blocked low concentrations of thrombin (1 nM) but were not effective at relatively high concentrations of the enzyme (30 nM). However, total inhibition of platelet aggregation was demonstrated by blocking both PAR-1 and PAR-4 but not either receptor alone. Limitations on the expression of PAR-1 on platelets of different species have hampered our attempts at understanding the significance of PAR-1 relative to other thrombin responsive mechanisms.

Several comparative studies demonstrated that among laboratory species, only guinea pig and nonhuman primates expressed PAR-1. The potential role for thrombin mediated PAR-1 activation in thrombus formation in vivo was first demonstrated with the use of a blocking antibody recognizing the hirudin-like domain in the amino terminal extension of PAR-1 in a non-human primate model of thrombosis. Interruption of thrombin's ability to bind to this region inhibited arterial thrombus formation measured by a decrease in cyclic flow reductions.

More recently, we have identified novel, selective PAR-1 antagonists, represented by RWJ-56110 and RWJ-58259, which effectively inhibit ex vivo platelet aggregation in guinea pigs. These agents represent excellent tools to probe the role of PAR-1 in normal and pathological states. RWJ-58259 as a prototype compound to assess the effects of PAR-1 inhibition in platelet dependent thrombosis in vivo. RWJ-58259 in two models of thrombosis, the arteriovenous shunt and photoactivation models. PAR-1 inhibition had no appreciable inhibitory effect on thrombus formation despite significant inhibition of both thrombin and SFLLRN-induced platelet aggregation.

Recently, studies exploring the interaction of PAR-3 and PAR-4, as they relate to murine platelet responsiveness to thrombin, were described. Using recombinant systems, PAR-3 was shown to act as a cofactor for thrombin-induced PAR-4 cleavage and activation through binding of thrombin to PAR-3's hirudin-like domain. This unique observation suggests a scenario whereby guinea pig platelets possess the potential for two equivalent thrombin-activating pathways:

PAR-1 and PAR-3/PAR-4. A PAR profile as such would limit the usefulness of this species as a model for assessing the role of human PAR-1 in platelet dependent disorders since human platelets do not express PAR-3. Thus, evaluation of the potential antithrombotic efficacy of a PAR-1 antagonist such as RWJ-58259 would require studies that overcome this species issue, such as the use of non-human primate models.

VASCULAR INJURY RESPONSES

Some researchs suggest that thrombin and, in particular, thrombin activation of PAR-1, plays a pivotal role in vascular injury responses associated with thrombosis, atherosclerosis, balloon angioplasty, and stent implantation. The identified vascular actions of thrombin-induced PAR-1 activation comprise the critical processes mediating vascular injury. Although there are many other mediators participating in the injury response, thrombin may act more globally to initiate these responses. Platelet adherence and aggregation, partly the result of PAR-1 activation by thrombin, are critical early events in initiating other vascular responses to injury through the release of other growth factors such as platelet derived growth factor.

Thrombin and PAR-1 mediated endothelial cell activation causes increased vascular permeability, release of nitric oxide, endothelin, prostaglandins, von Willebrand factor, and increased adhesion molecule expression. The latter can lead to neutrophil adhesion at the site of injury. Thrombin-induced vasoconstriction is PAR-1 mediated. Acutely, this vasoconstriction may result in underperfusion or ischemia due to narrowing of the vessel lumen and potentially increasing the risk of an occlusive thrombus. Vascular smooth muscle cell proliferation in rat, rabbit, and human is induced by PAR-1 activation.

In addition to cellular proliferation, extracellular matrix protein turnover is an important component of the vascular injury response particularly in mediating restenosis following angioplasty in human vessels. Thrombin activation of PAR-1 causes vascular smooth muscle procollagen synthesis, increased synthesis and activation of matrix metalloproteinases in endothelial cells, and endothelial cell secretion of extracellular matrix proteins. The local elevation of thrombin concentrations at a site of thrombus is well established. The significant incidence of intramural thrombus following angioplasty of atherosclerotic vessels may provide a sustained exposure to thrombin at the vascular injury site.

In addition, PAR-1 protein expression and mRNA appear to be upregulated in vascular injury in rat and baboon arteries, where PAR-1 expression is associated with proliferating vascular smooth muscle cells. Similar PAR-1 upregulation has been demonstrated following vascular injury in rabbit and mouse carotid arteries and in vein graft remodeling in rats. PAR-1 expression is also increased in human atherosclerotic vessels. Thus, the association of elevated thrombin concentrations and increased PAR-1 expression at a site of vascular injury is consistent with a role for thrombin and PAR-1 in vascular injury.

Inhibitors of thrombin, such as hirudin, reduce neointimal thickening induced by balloon angioplasty in normal and hypercholesterolemic rabbits and pigs, and in normal rats. However, thrombin inhibitors will not only inhibit the activation of PAR-1 but will also inhibit activation of other thrombin receptors such as PAR-3 and PAR-4 and will prevent the other enzymatic actions of thrombin such as coagulation and fibrin polymerization. Thus, the effectiveness of the direct thrombin inhibitors does not prove a role for PAR-1 in vascular injury.

Therefore, a number of alternative approaches have been used to assess the specific role of PAR-1. For example, the vascular injury response in mice made genetically deficient in PAR-1 is altered compared to wild type mice. Specifically, intimal area and increased medial area in response to injury tended to be reduced in PAR-1 deficient mice. However, there appeared to be more significant changes in remodeling in response to injury. Although vessel and lumen diameter tended to increase in wild type mice following injury, vessel diameter was unchanged and lumen diameter actually decreased in PAR-1 deficient mice despite the reduced neointimal thickening. This suggests a role for PAR-1 in positive remodeling following vascular injury.

Cell density in the neointima was also greater in PAR-1 deficient mice. The increased cell density in normal and injured vessels may have resulted from reduced matrix synthesis and secretion as a result of absence of PAR-1 activation. An antisense oligodeoxynucleotide to PAR-1 has been used to assess the specific role of PAR-1 in a vascular injury model in rabbits. This oligonucleotide inhibited expression of PAR-1 mRNA in vivo and inhibited the proliferative response to thrombin in vascular smooth muscle cells. However, the degree of intimal thickening was not altered. In the same model, hirudin significantly reduced the intimal thickening response.

INFLAMMATION AND TISSUE INJURY

Thrombin contributes to many of the inflammatory and reparative processes accompanying tissue injury including changes in vascular permeability, leukocyte-endothelial interactions, and connective tissue remodeling. Through activation of PAR-1, thrombin mediates several important effects on the vascular endothelium such as stimulation of the synthesis and release of various cytokines and growth factors and induction of adhesion molecule expression. These soluble mediators, which include interleukin-1, interleukin-6, interleukin-8, and monocyte chemoattractant protein-1, act as chemoattractants for and activators of peripheral blood leukocytes.

PAR-1 dependent increases in vascular contraction and vascular permeability in vitro also indicate a direct effect of thrombin on the vascular response associated with tissue injury. Taken together, these many actions of thrombin on the vascular endothelium indicate a coordinated process that facilitates the transmigration of leukocytes from the blood to the extravascular space.

In addition to thrombin's role in mediating a proinflammatory response during tissue injury, thrombin exerts a variety of additional actions on cells of the vascular and extravascular space. Thrombin exhibits in vitro angiogenic activity through induction of endothelial tube formation in MatrigelTM as well as promoting neovascularization in vivo. While this thrombin-dependent angiogenic response is important for normal tissue healing, it may contribute to pathological conditions such as in tumor angiogenesis and atherosclerotic plaque formation.

The extravascular actions of thrombin include activation of tissue fibroblasts, monocytes/macrophages, and mast cells. Fibroblasts are important participants during tissue repair by elaborating a variety of cytokines and growth factors for both autocrine and paracrine actions. Secretion of vascular endothelial growth factor, a key angiogenic factor, pro-fibrotic connective tissue growth factor, and the inflammatory cytokine interleukin-8 highlight the importance of PAR-1-induced fibroblast activation in tissue injury. In addition to its direct effect on the release of soluble mediators from fibroblasts, thrombin synergizes with interleukin-1β to secrete prostaglandin E2 from fibroblasts. This interaction is associated with both increased vascular permeability and vasorelaxation, which are contributing factors to tissue edema. The effects on fibroblast secretagogue activity are accompanied by thrombin's strong mitogenic activity toward fibroblasts.

The importance of PAR-1 as the primary thrombin receptor mediating these inflammatory associated events has been explored further using several in vivo models. PAR-1 deficient mice provide a valuable model to explore the direct actions of thrombin on cell and tissue responses. PAR-1 activation by either thrombin or a PAR-1 agonist peptide increased vascular permeability and vasoconstriction in the pulmonary microvasculature. These changes were absent in lung preparations from PAR-1 deficient mice, consistent with the expression of functional PAR-1 on the vasculature.

In a model of inflammatory paw edema, PAR-1 expressing mice challenged with either thrombin or a PAR-1 agonist peptide exhibited increased plasma extravasation which was attenuated by either the neurokinin 1 receptor antagonist SR140333 or ablation of sensory nerves by capsaicin. Mice deficient in PAR-1 were protected from the increased edema and leukocyte infiltration confirming that PAR-1 mediates this inflammatory response and does so via a neurogenic mechanism. In a model of rat paw edema, the thrombin inhibitor HirulogTM attenuated carrageenin-stimulated paw edema, implicating thrombin as a central mediator of the response.

Degranulation of mast cells by compound 48/80 attenuated the responses to PAR-1 activation, implicating mast cells, either directly or indirectly, as key participants in the inflammatory response. Mast cell granule constituents play important roles in immediate (type I) hypersensitivity, inflammation, tissue remodeling, and angiogenesis. Upon activation, mast cells secrete a range of soluble mediators including histamine, heparin, leukotrienes, cytokines, growth factors, and neutral proteases.

Thus, mast cells represent an efficient system through which powerful mediating factors and enzymes are deposited locally upon activation. The presence of PAR-1 on mast cells in various normal human tissues using immunohistochemical techniques. PAR-1 was distributed on the plasma membrane and on the membranes of the intracellular tryptase-positive granules of the mast cells. Such findings are consistent with earlier observations demonstrating that thrombin can induce mast cell activation.

More recently, PAR-1 mRNA has also been identified in mast cells. Taken together, these observations indicate the presence of mast cell receptors for thrombin that initiate or potentiate mast cell activation. Therefore, the function of PAR-1 on the membranes of intracellular granules may be to sustain further mast cell degranulation by replenishing internalized PAR-1 from the plasma membrane upon activation. The discovery of a functional PAR-1 on mast cells suggests a novel mechanism to control mast cell activity through the regulation of PAR-1.

TUMOR MICROENVIRONMENT (TME)

Of the many factors secreted by tumor cells, thrombin has been correlated with the stage and type of carcinoma and is associated with cell invasion and extracellular matrix degradation. Increased proteolytic activity in the TME can favor capillary sprout elongation and lumen formation during angiogenesis. The numerous cellular actions of thrombin on platelets, vascular cells, mast cells, and leukocytes previously noted may contribute to tumor invasion. Additionally, thrombin-mediated connective tissue deposition and development of tissue fibrosis, characteristics of normal wound healing, are processes also associated with cellular metastasis.

In vitro studies have demonstrated increased tumor cell adhesion to endothelium, extracellular matrix and platelets, enhanced metastatic capacity of tumor cells, and activated cell growth and stimulation of angiogenesis in response to thrombin and PAR-1 agonist peptides. PAR-1 has been localized in pancreas tumor cells, carcinoma, and melanoma cell lines. In breast carcinoma cells, the level of PAR-1 expression has been correlated with the degree of invasiveness.

Furthermore, B16F10 melanoma cells transfected with PAR-1, enhanced thrombin-induced tumor cell adhesion to fibronectin 2.5-fold and pulmonary metastasis as high as 39-fold compared to untransfected thrombin-treated tumor cells. The expression of PAR-1 in malignant and benign human tumor tissues was recently described in the context of the surrounding cell types in the TME using immunohistochemistry and in situ hybridization.

PAR-1 was expressed not only on the malignant carcinoma cells, but also on the cell types forming the TME such as the local macrophages, mast cells, reactive stromal fibroblasts, vascular endothelial, and smooth muscle cells. The presence of PAR-1 in these cell types in the TME suggests a role in the actions of thrombin in cell metastasis. The interface between the invading malignant cells and the hosting stromal cells, referred to as the TME, possesses an array of well-orchestrated cell signaling mechanisms which facilitate tumor invasion by stimulating tumor cell proliferation and degradation and remodeling of the extracellular matrix.

While PAR-1 expression has previously been shown on endothelial cells, vascular smooth muscle cells and mast cells, little is known about the presence of PAR-1 on macrophages. It has been reported that macrophages can secrete thrombin and that thrombin is localized in pulmonary alveolar macrophages. It is possible that PAR-1 activation might provide a stimulus for macrophages to proliferate, migrate and/or phagocytize degraded stromal proteins, in addition to synthesizing and secreting thrombin into the TME.

PAR-1 was also identified on the reactive stromal fibroblasts surrounding the metastatic tumor cells, which were also positive for smooth muscle actin and DNA topoisomerase IIβ, a proliferation marker. PAR-1 immunoreactivity was not detected on the stromal fibroblasts surrounding the benign, non-metastatic, or normal epithelial cells, which were smooth muscle actin and topoisomerase IIβ negative in the tissues analyzed. Reactive stromal fibroblasts, which exhibited characteristics of myofibroblasts, are frequently associated with cancers of epithelial origin: a process known as desmoplasia.

These cells appear to influence the invasive and metastatic potential of carcinoma cells by an unidentified mechanism. The role of thrombin on cell function is multi-faceted. Induction of the coagulation pathway on the cell surface of platelets, macrophages, vascular cells, or tumor cells in response to tissue injury or inflammation, provides a means by which thrombin is generated and becomes accessible not only intravascularly but also to the resident cells in the subendothelial tissue. The generation of thrombin in response to tissue injury and inflammatory processes results in the coordinated activation of cells within the vascular and extravascular compartments. The identification of a novel proteolytically activated receptor for thrombin, PAR-1, and evidence for its ubiquitous expression on numerous cells types has provided an explanation for thrombin's cellular actions.

Accumulating evidence indicates that PAR-1 does indeed mediate many of the cellular actions of thrombin. This includes activation, adhesion, and aggregation of platelets in arterial thrombosis. Within the vascular milieu, thrombin can modulate the interaction of leukocytes and endothelium to initiate a cascade of events leading to extravasation of cells and protein to the extravascular space, important components of inflammation and edema. Thrombin can also initiate or potentiate mast cell activation. In addition, thrombin can stimulate vascular smooth muscle cell and fibroblast proliferation as well as matrix remodeling, critical components of the vascular injury response. The upregulation of PAR-1 in fibroblasts surrounding malignant tumors suggests a possible role in tumor metastasis.

REFERENCES

Ambrose, E. J., and Easty, D. M., *Cell Biology: Reading*, Mass., Addison-Wesley, 1970.

Baker, J. J. W., and Allen, G. E., *The Study of Biology* [2nd ed.], Reading, Mass., Addison-Wesley, 1971.

Dowben, R. M., *Cell Biology*, New York, Harper & Row, 1971.

4

Cell Signalling and Communication

Cell signalling is part of a complex system of communication that governs basic cellular activities and coordinates cell actions. The ability of cells to perceive and correctly respond to their microenvironment is the basis of development, tissue repair, and immunity as well as normal tissue homeostasis. Errors in cellular information processing are responsible for diseases such as cancer, autoimmunity, and diabetes. By understanding cell signalling, diseases may be treated effectively and, theoretically, artificial tissues may be yielded.

Traditional work in biology has focused on studying individual parts of cell signalling pathways. Systems biology research helps us to understand the underlying structure of cell signalling networks and how changes in these networks may affect the transmission and flow of information. Such networks are complex systems in their organization and may exhibit a number of emergent properties including bistability and ultrasensitivity. Analysis of cell signalling networks requires a combination of experimental and theoretical approaches including the development and analysis of simulations and modelling.

The external signall may enter a cell via four major pathways:

— Hydrophobic molecules.

— Ion channels.
— G-protein-coupled receptors.
— Enzymes.

In many cases, the signal continues to propagate within the cell and often reaches nuclear DNA to express proteins.

HYDROPHOBIC MOLECULES

Hydrophobic molecules can move in and out of cells by passing through lipid bilayers. Unlike most signalling cascades which occur within the same cell, newly generated NO or arachidonic acid diffuses to act on target molecules in neighboring cells.

Nitric Oxide

Nitric oxide (NO) is produced from the following reaction:

$$\text{L-arginine} \xrightarrow{\text{NOS}} \text{L-citrulline} + \text{NO}$$

where NOS represents NO synthase. After NO is generated in a cell, it diffuses to act on target molecules in neighboring cells. NO may stimulate soluble guanylyl cyclase to produce cGMP which regulates several enzymes and ion channels. In smooth muscle, an important action of cGMP is to induce muscle relaxation. Normally, cGMP will soon be converted into GMP by phosphodiesterase. The well known drug for impotence, Viagra (sildenafil citrate), inhibits phosphodiesterase.

Arachidonic Acid

Arachidonic acid is generated from phospholipid hydrolysis catalyzed by phospholipase. Like NO, its target is usually not located in the cell where it was first generated. After diffusing to target cells, arachidonic acid may activate protein kinase, resulting in phosphorylation of target molecules. Many of its target molecules are involved in learning and other neuronal activities.

Steroids

The major role of steroids is to regulate transcription, since many steroid receptors are transcription factors. Steroid-bound SR may recruit SRC (steroid receptor coactivator) and histone acetyltransferases to stimulate transcription of the target gene. In the absence of the steroid, several steroid receptors may recruit corepressors (e.g., SMRT and NCoR) and histone deacetylases to repress transcription.

ION CHANNELS

Ion channels are membrane proteins that allow ions to pass through. In terms of ion selectivity, they are classified as calcium channels, sodium channels, potassium channels, etc. In terms of gating (how channels are opened), they may be classified as voltage-gated channels, ligand-gated channels, etc.

The major role of sodium and potassium ions is to regulate membrane potential. However, the calcium ion also plays important roles in other cellular functions, since many enzymes are calcium-dependent. There are two pathways that Ca^{2+} may enter the cytosol.

— Through voltage-activated Ca^{2+} channels whose opening probability depends on the membrane potential.

— Through IP_3-sensitive Ca^{2+} channels whose opening probability is regulated by IP_3 (inositol trisphosphate).

The opening of Ca^{2+} stores in the endoplasmic reticulum may result in calcium waves. Calmodulin (structure at Protein Data Bank) is an important molecule that can bind to Ca^{2+} to modulate its activity.

G-PROTEIN-COUPLED RECEPTORS

The major role of G-protein-coupled receptors is to transmit signals into the cell. They are characterized by seven transmembrane segments. This class of membrane proteins can respond to a wide range of agonists, including photon, amines, hormones, neurotransmitters and proteins. Some agonists bind to the extracellular loops of the receptor, others may penetrate into the transmembrane region.

- Before agonist binding to the G-protein-coupled receptor, the three subunits of G protein are bound together.
- The agonist binding causes interaction between the G-protein-coupled receptor and the G protein.
- Their interaction results in the dissociation between a and bg subunits of the G protein. The separated a and/or bg subunits may then interact with effectors.

G Proteins

The full name of "G protein" is GTP-binding protein because in the active state it binds to GTP (guanosine triphosphate). There are two types of G proteins: heterotrimeric G proteins and monomeric G proteins (or small G proteins). G-protein-coupled receptors are coupled to heterotrimeric G proteins.

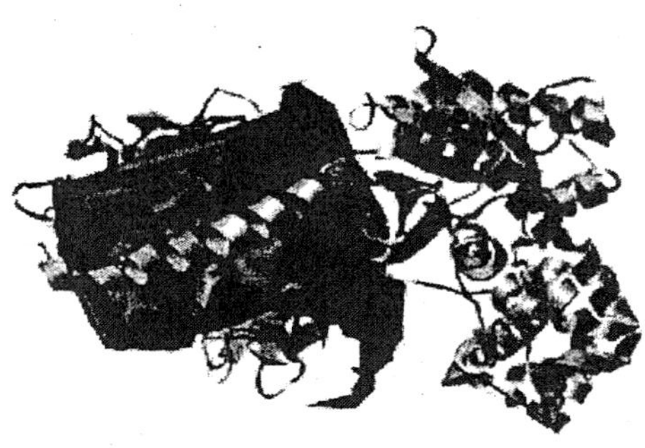

Figure 1. The structure of heterotrimeric G protein

The heterotrimeric G protein consists of three subunits: α, β and γ. Based on the differences in their genes, 20 α, 6 β and 12 γ subunits have been identified. Their molecular weights are in the following ranges:

— α subunit: 39 - 46 kD

— β subunit: 35 - 39 kD

— γ subunit: ~ 8 kD

In the inactive state, the a subunit binds to GDP and the three subunits are attached together. When the a subunit binds to GTP, its affinity to the bg subunits is decreased, resulting in their dissociation. The separated a and/or bg subunits can then interact with their effectors.

Cycling of G protein

The rate of GTP hydrolysis by Ga is regulated by a special class of proteins.

Effectors are the target molecules of G protein a or bg subunit. The bg subunits may act on adenylyl cyclase, phospholipase A2, phospholipase C, ion channels, calcium ATPase etc.

Adenylyl cyclase

Adenylyl cylase (also known as "adenylate cyclase") catalyzes the conversion of ATP to cAMP, which is an important "second messenger"

Phospholipase A2

Phospholipase A2 (PLA2) cleaves phospholipids, generating arachidonic acid directly.

Phospholipase C

Phospholipase C (PLC) also cleaves phospholipids, but at a different site than PLA2, thereby generating diacylglycerol (DAG). There are three types of PLC: β, γ, and δ. The effector of G protein a and βγ subunits is PLC-β.

It generates DAG and IP_3 (inositol 1,4,5-trisphosphate). The latter can activate an IP_3-sensitive Ca^{2+} channel in endoplasmic reticulum.

Second Messengers

Second messengers are the signalling molecules generated by the stimulation of cell-surface receptors. For example, cAMP, arachidonic acid, DAG and IP_3 generated by the activation of G-protein-coupled receptors are second messengers. The agonists which activate G-protein-coupled receptors are first messengers.

CELL-SURFACE ENZYMES

Signalling enzymes which are cell-surface receptors:
- Receptor tyrosine kinases
- Receptor serine/threonine kinases
- Receptor tyrosine phosphatase

catalyze the dephosphorylation of the tyrosine residue.

INTRACELLULAR SIGNALLING

- cAMP
- NF-kB
- STAT
- MAPK

The targets of cAMP include:
- cAMP-dependent protein kinase (PKA).
- Cyclic nucleotide gated ion channels.
- Guanine exchanging factors Epac1 and Epac2 that regulate the activity of Rap1.

PKA consists of two catalytic subunits and two regulatory subunits. cAMP binds to two sites on each regulatory subunit. Binding of four cAMP molecules causes the release of free and active catalytic subunits, which may phosphorylate serine and threonine residues on target proteins. The active subunit phosphorylates a CREB protein, resulting in transcription.

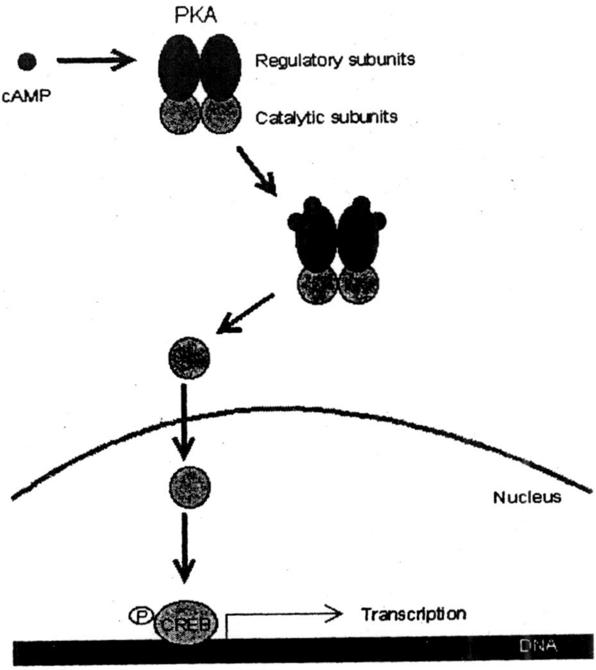

Figure 2. A cAMP/PKA signalling pathway.

Signalling by NF-kB: NF-kB can be activated by a variety of stimuli, including TNF-α (tumor necrosis factor α), interleukin 1 (IL-1), T and B cell mitogens, bacterial lipopolysaccharide (LPS) and viral proteins. Current experimental results suggest that some protein kinase in the MAPKKK group is involved, such as NIK (NF-kB-inducing kinase).

— Binding of TNF-a to its receptor recruits TRADD (TNF receptor-associated death domain protein), RIP (receptor-interacting protein) and TRAF2 (TNF receptor-associated factor 2).
— TRAF2 activates IkB kinase (IKK).
— The active IKK activates NF-kB by degrading the inhibitor protein IkB.
— After NF-kB is activated, it may serve as a transcription factor or inhibit apoptosis.

Signalling by STAT: STAT stands for "Signal Transducer and Activator of Transcription", because this class of proteins function as both signal transducer and transcription activator. They are inactive as a monomer. Activation involves phosphorylation and dimerization.

The DNA binding domain may be used to activate transcription and the SH2 domain may receive signal from tyrosine kinase such as Jak (a non-receptor tyrosine kinase). Most cytokines use the Jak-STAT signalling pathway.

The MAPK Signalling Pathway: The Mitogen-Activated Protein Kinase (MAPK) pathways transduce a large variety of external signals, leading to a wide range of cellular responses, including growth, differentiation, inflammation and apoptosis. These pathways are characterized by the following general path:

$$\text{Stimulus} \rightarrow \text{MAPKKK} \rightarrow \text{MAPKK} \rightarrow \text{MAPK} \rightarrow \text{Response}$$

where MAPKK is the kinase of MAPK and MAPKKK is the kinase of MAPKK. In most cases, the MAPKKK is activated by small G proteins such as Ras, Rac and Rap1. However, it may also be activated by other enzymes. In mammals, three major MAPK pathways have been identified: MAPK/ERK, SAPK/JNK, and p38 MAPK.

APOPTOSIS

Apoptosis is a tightly regulated form of cell death, also called the programmed cell death. Morphologically, it is characterized by chromatin condensation and cell shrinkage in the early stage. Then the nucleus and cytoplasm fragment, forming membrane-bound apoptotic bodies which can be engulfed by phagocytes. In contrast, cells undergo another form of cell death, necrosis, swell and rupture. The released intracellular contents can damage surrounding cells and often cause inflammation. Apoptosis is an important process during normal development. It is also involved in aging and various diseases such as cancer, AIDS, Alzheimer's disease and Parkinson's disease.

During apoptosis, the cell is killed by a class of proteases called caspases. More than 10 caspases (web link) have been identified. Some of them (e.g., caspase 8 and 10) are involved in the initiation of apoptosis, others (caspase 3, 6, and 7) execute the death order by destroying essential proteins in the cell. The apoptotic process can be summarized as follows:

— Activation of initiating caspases by specific signals.

— Activation of executing caspases by the initiating caspases which can cleave inactive caspases at specific sites.

— Degradation of essential cellular proteins by the executing caspases with their protease activity.

— *Death receptors*: Fas/CD95, DR4/DR5, DR3, and TNFR (Tumor Necrosis Factor Receptor).

— *Adaptors*: FADD (Fas-associated death domain protein) and TRADD (TNFR-associated death domain protein).

— *Activation*: Binding of death ligands (FasL/CD95L, TRAIL/APO-2L, APO-3L and TNF) induces trimerization of their receptors, which then recruit adaptors and activate caspases.

— *Note*: TRADD is involved only in the coupling between caspases and DR3 or TNFR. This adaptor can also recruit other proteins (web link) to inhibit apoptosis through the NF-kB pathway.

A variety of death ligands (FasL/CD95L, TRAIL, APO-3L and TNF) can induce apoptosis. It is natural to see if they can kill cancer cells without affecting normal cells. TNF was first investigated in the 1980s for cancer therapy, but with disappointing results. Then CD95L (FasL) was tested in the 1990s. The results were still not satisfactory. Recently, TRAIL has been demonstrated to be highly selective for transformed cells, with minimal effects on normal cells. It could be an effective drug for both cancer and AIDS.

TRAIL

TRAIL receptors can be divided into two categories:

— *Death receptors*: TRAIL-R1 and TRAIL-R2 contain the death domain, capable of inducing apoptosis.

— *Decoy receptors*: TRAIL-R3 and TRAIL-R5 lack the death domain while TRAIL-R4 contains a truncated non-functional death domain. These three receptors can bind to TRAIL, but cannot induce apoptosis.

TRAIL-R5 is secreted to the extracellular fluid. All other receptors are transmembrane proteins. TRAIL (TNF-related apoptosis-inducing ligand) is a protein consisting of 281 amino acids. It is also called APO-2L. Five TRAIL receptors have been identified. Two of them can lead to apoptosis. The other three serve as "decoy" receptors. A cell expressing more decoy receptors is more likely to survive upon binding of the death ligands.

TRAIL as Cancer Therapeutic

TRAIL has been demonstrated to kill a wide variety of tumor cells with minimal effects on normal cells. This is because TRAIL's death receptors (TRAIL-R1 and TRAIL-R2) are mainly expressed in transformed cells while its decoy receptors (TRAIL-R3, TRAIL-R4 and TRAIL-R5) are expressed in normal cells. The major concern about using TRAIL as a therapeutic is its toxicity to human hepatocytes. TRAIL can cause significant death of normal human hepatocytes, but has no effect on the normal hepatocytes of mice. In the future, as we have better understanding on the regulation of receptor expression and signalling cascades, it may be possible to save normal cells from the toxic effects of TRAIL.

TRAIL and AIDS

AIDS (acquired immune deficiency syndrome) is characterized by the depletion of T cells. It has been demonstrated that HIV can induce apoptosis in both infected cells and their neighboring (bystander) cells. Recently, it was further shown that those bystander cells were killed mainly through the TRAIL-mediated apoptosis. Experimental results suggest that the extracellular Tat protein, released from HIV-infected cells, can induce production of TRAIL by macrophages, resulting in the apoptosis of bystander T cells.

TRAIL has minimal effects on normal cells (except for human hepatocytes). Presumably, the regulation of TRAIL receptor expression in bystander cells should be altered, but the detailed mechanism is not known. A recent report indicates that the HIV's surface protein, gp120, is involved. The most challenging question in AIDS therapy is how to eradicate the HIV that persists in "latent reservoirs" (e.g., resting memory T cells) even after prolonged antiretroviral therapy.

Recently, it was shown that TRAIL can kill the latent cells. Since TRAIL also kills the transformed bystander cells, it is recommended that TRAIL be given to patients after prolonged HAART (highly active antiretroviral therapy). At this time, only the latent cells remain and their neighboring cells may not be susceptible to TRAIL.

Mitochondria, Apoptosis and Aging

Two major mechanisms are involved in the aging process: telomere shortening and free radical attacks. Mitochondria are the primary source of free radicals due to inevitable leakage of electrons from the electron transport chain. Since free radicals are highly reactive, they may interact with DNA and proteins to alter cellular functions. A very destructive target is the voltage-dependent anion channel (VDAC) which plays the central role in mitochondria-mediated apoptosis.

As mentioned in the p53 page, the release of cytochrome c from mitochondria into the cytoplasm, in concert with Apaf1, can activate caspase 9, which then activate caspase 3 to execute the death program. Cytochrome c and other small molecules may pass through the mitochondrial permeability transition pore which comprises VDAC in the outer membrane, the adenine nucleotide translocator (ANT) in the inner membrane, and several auxiliary proteins that include the Bcl-2 family involved in apoptosis.

Bcl-2 family of proteins may be divided into three groups:

— *Anti-apoptotic*: They share sequence homology at BH1, BH2, BH3, and BH4 domains (BH = Bcl-2 homology). Examples: Bcl-2 itself and Bcl-xL.

— *Pro-apoptotic*: They share sequence homology at BH1, BH2 and BH3 domains. Examples: Bax and Bak

— *BH3-only proteins*: They are pro-apoptotic, but share homology only at the BH3 domain. Examples: Bid, Bik, and Bim.

It has been shown that the BH4 domain is required for anti-apoptotic activity and that the BH3 domain is essential and sufficient for pro-apoptotic activity. Both Bcl-2/Bcl-xL and Bax/Bak interact with VDAC (not ANT) to change the membrane permeability. The anti-apoptotic Bcl-2 and Bcl-xL close the VDAC channel. The pro-apoptotic Bax and Bak can open the VDAC channel to release cytochrome c. The action of the BH3-only proteins is not clear.

Free radicals such as superoxide (O_2^+) can also open the VDAC channel, resulting in the release of cytochrome c and cell death. This could be a major mechanism for the aging process. Several studies have documented a significant loss of cardiac and skeletal myocytes during normal aging.

Free radicals and telomer shortening may both contribute to the aging process in the following way. The free radicals cause cell death. Although new cells may be generated by cell division, the telomer will become shorter in the daughter cells. As the telomer length decreases substantially, it will activate p53 to stop the cell division. Thus, aged cells will no longer produce new cells.

PRINCIPLES OF CELL SIGNALLING

Information can come in a variety of forms, and communication frequently involves converting information signals from one form to another.

When you phone a friend, for instance, the sound waves of your voice are converted to electrical signals that travel over a telephone wire. The critical points in this relay occur where the message is converted from one form to another. This process of conversion is called signal transduction (Figure 3). The signals that pass between cells are far simpler than the sorts of messages that humans ordinarily exchange. In a typical communication between cells, the signalling cell produces a particular type of signal molecule that is detected by the target cell.

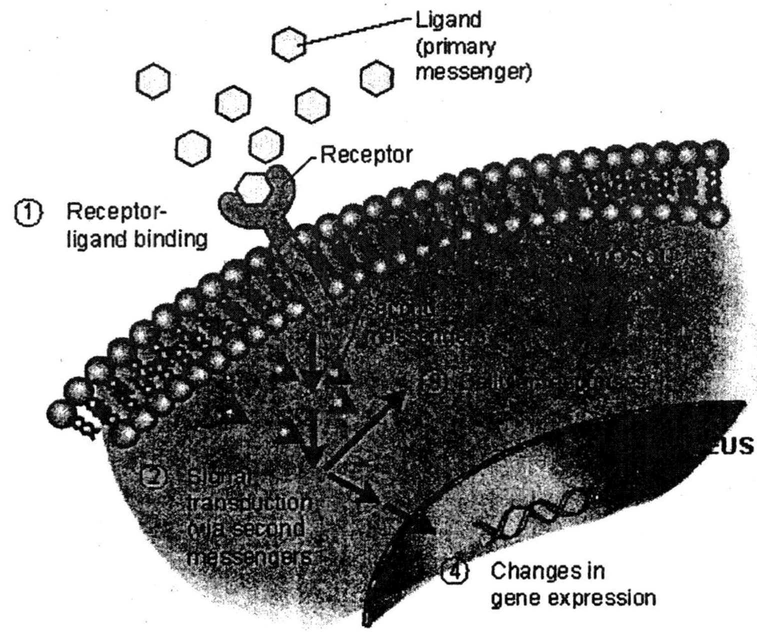

Figure 3. Signal transduction

The target cells possess receptor proteins that recognize and respond specifically to the signal molecule. Signal transduction begins when the receptor protein on the target cell receives an incoming extracellular signal and converts it to the intracellular signals that direct cell behavior. Single cells and cells in multicellular organisms use hundreds of kinds of extracellular molecules to send signals to one another-proteins, peptides, amino acids, nucleotides, steroids, fatty acid derivatives, and even dissolved gases-but they rely on only a handful of basic styles of communication for getting the message across. In multicellular organisms, the most "public" style of communication involves broadcasting the signal throughout the whole body by secreting it into the bloodstream (in an animal) or the sap (in a plant).

Signal molecules used in this way are called hormones, and in animals, the cells that produce hormones are called endocrine cells. For example, part of the pancreas is an endocrine gland that produces the hormone insulin, which regulates glucose uptake in cells all over the body. Somewhat less public is the process known as paracrine signalling. In this case, rather than entering the bloodstream, the signal molecules diffuse locally through the extracellular medium, remaining in the neighborhood of the cell that secretes them. Thus they act as local mediators on nearby cells.

Many of the signal molecules that regulate inflammation at the site of an infection or control cell proliferation in a healing wound function in this way. Neuronal signalling constitutes a third form of cell communication. Like endocrine cells, neurons can deliver messages across long distances. In the case of neuronal signalling, however, a message is not broadcast widely but is delivered quickly and specifically to individual target cells through private lines. The axon of a neuron terminates at specialized junctions (synapses) on target cells that can lie far from the neuronal cell body. The axons that connect a person's spinal cord and big toe, for example, can be more than 1 m in length.

When activated by signals from the environment or from other nerve cells, a neuron sends electrical impulses racing along its axon at speeds of up to 100 m/sec. On reaching the axon terminal, these electrical signals are converted into a chemical form: each electrical impulse stimulates the nerve terminal to release a pulse of an extracellular chemical signal called a neurotransmitter. These neurotransmitters then diffuse across the narrow (< 100 nm) gap between the axon-terminal membrane and the membrane of the target cell in less than 1 msec. A fourth style of signal-mediated cell-cell communication-the most intimate and short-range of all-does not require the release of a secreted molecule.

Instead, the cells make direct contact through signalling molecules lodged in their plasma membranes. The message is delivered when a signal molecule anchored in the plasma membrane of the signalling cell binds to a receptor molecule embedded in the plasma membrane of the target cell. In embryonic development, for example, such contact-dependent signalling plays an important part in tissues in which adjacent cells that are initially similar are destined to become specialized in different ways. To relate these different signalling styles, imagine trying to advertise a potentially stimulating lecture-or a concert or football game.

An endocrine signal would be akin to broadcasting the information over a radio station. A flyer posted on select notice boards would be the equivalent of a localized paracrine signal. Neuronal signals-long-distance but personal-would be similar to a phone call or an e-mail, and contact-dependent signalling would be like a good, old-fashioned face-toface conversation.

Cell Responds to Signals

A typical cell in a multicellular organism is exposed to hundreds of different signal molecules in its environment. These may be free in the extracellular fluid, embedded in the extracellular matrix in which cells rest, or bound to the surfaces of neighboring cells. Each cell must respond selectively to this mixture of signals, disregarding some and reacting to others, according to the cells' specialized function. Whether a cell responds to a signal molecule depends first of all on whether it possesses a receptor for that signal. Without the appropriate receptor, a cell will be deaf to the signal and will not react. By producing only a limited set of receptors out of the thousands that are possible, the cell restricts the range of signals that can affect it. But this limited range of signals can still be used to control the behavior of the cell in complex ways. The complexity is of two sorts.

First, one signal, binding to one type of receptor protein, can cause a multitude of effects in the target cell: it can alter the cell's shape, movement, metabolism, and gene expression. As we shall see, the signal from a cell-surface receptor is generally conveyed into the cell interior via a set of interacting molecular mediators that are capable of producing widespread effects in the cell. Furthermore, this intracellular relay system and the intracellular targets on which it acts vary from one type of specialized cell to another, so that different types of cells respond to the same signal in different ways. For example, when a heart muscle cell is exposed to the neurotransmitter acetylcholine, the rate and force of its contractions decrease, but when a salivary gland is exposed to the same signal, it secretes components of saliva. These responses occur rapidly-within seconds to minutes-because the signal affects the activity of proteins and other molecules that are already present inside the cells.

The second kind of complexity arises because a typical cell possesses a whole collection of different receptors-tens to hundreds of thousands of receptors of a few dozen types. Such variety makes the cell simultaneously sensitive to many extracellular signals. These signals, by acting together, can evoke responses that are more than just the sum of the effects that each signal would evoke on its own. The intracellular relay systems for the different signals interact, so that the presence of one signal modifies the responses to another. Thus one combination of signals may simply enable a cell to survive; another may drive it to differentiate in some specialized way; and another may cause it to divide.

In the absence of any signals, most animal cells are programmed to kill themselves. Because the execution of such a complex programme often requires the synthesis of new proteins, it may take the cell hours to fully respond to the incoming signals. Overall, the integration of extracellular cues allows a relatively small number of signal molecules, used in different combinations, to exert subtle and complex control over cell behavior.

Intracellular Signalling Pathways

Signal reception begins at the point where a signal originating outside the target cell encounters a target molecule belonging to the cell itself. In virtually every case, the target molecule is a receptor protein (also called a receptor), and each receptor is usually activated by only one type of signal. The receptor protein performs the primary transduction step: it receives an external signal and generates a new intracellular signal in response.

As a rule, this is only the first event in a chain of intracellular signal transduction processes. In this game of molecular tag, the message is passed from one intracellular signalling molecule to another, each activating or generating the next signalling molecule in line, until, say, a metabolic enzyme is kicked into action, a gene is switched on, or the cytoskeleton is tweaked into a new configuration. This final outcome is called the response of the cell. These relay chains, or signalling cascades, of intracellular signalling molecules have several crucial functions:

1. They transform, or transduce, the signal into a molecular form suitable for passing the signal along or stimulating a response.
2. They relay the signal from the point in the cell at which it is received to the point at which the response is produced.
3. In many cases, signalling cascades also amplify the signal received, making it stronger, so that a few extracellular signal molecules are enough to evoke a large intracellular response.
4. The signalling cascades can also distribute the signal so as to influence several processes in parallel: at any step

in the pathway, the signal can diverge and be relayed to a number of different intracellular targets, creating branches in the information flow diagram and evoking a complex response.

5. Each step in this signalling cascade is open to modulation by other factors, including other external signals, so that the effects of the signal can be tailored to the conditions prevailing inside or outside the cell.

Most signalling pathways trace a long and branching route, and enlist many molecular players, as they relay information from receptors at the cell surface to appropriate machinery in the cell's interior.

Plasma Membrane and Intracellular Enzymes

Extracellular signal molecules in general fall into two classes. The first and largest class of signals consists of molecules that are too large or too hydrophilic to cross the plasma membrane of the target cell. They rely on receptors on the surface of the target cell to relay their message across the membrane. The second, and smaller, class of signals consists of molecules that are small enough or hydrophobic enough to slip easily through the plasma membrane.Once inside, these signal molecules either activate intracellular enzymes or bind to intracellular receptor proteins that regulate gene expression.

For an extracellular signal to alter a cell within a few seconds or minutes, direct activation of an enzyme is an effective strategy. Nitric oxide (NO) is just such a signal. This dissolved gas diffuses readily out of the cell that generates it and enters neighboring cells. NO is made from the amino acid arginine and operates as a local mediator in many tissues. The gas acts only locally because it is quickly converted to nitrates and nitrites (with a half-life of about 5-10 seconds) by reaction with oxygen and water outside cells. Endothelial cells-the flattened cells that line every blood vessel-release NO in response to stimulation by nerve endings. This NO signal causes smooth muscle cells in the vessel wall to relax, allowing the vessel to dilate, so that blood flows through it more freely.

The effect of NO on blood vessels accounts for the action of nitroglycerine, which has been used for almost 100 years to treat patients with angina (pain caused by inadequate blood flow to the heart muscle). In the body, nitroglycerine is converted to NO, which relaxes coronary blood vessels and increases blood flow to the heart. Many nerve cells also use NO to signal neighboring cells: NO released by nerve terminals in the penis, for instance, triggers the local blood-vessel dilation that is responsible for penile erection. Inside many target cells, NO binds to the enzyme guanylyl cyclase, stimulating the formation of cyclic GMP from the nucleotide GTP.

Cyclic GMP itself is a small intracellular signalling molecule that forms the next link in the signalling chain that leads to the cell's ultimate response. The impotence drug Viagra enhances penile erection by blocking the degradation of cyclic GMP, prolonging the NO signal. Cyclic GMP is very similar in its structure and mechanism of action to cyclic AMP.

Bind to Intracellular Receptors

Gases such as NO are not the only signal molecules that can cross the plasma membrane. Hydrophobic signal molecules such as the steroid hormones-including cortisol, estradiol, and testosterone-and the thyroid hormones such as thyroxine all pass through the plasma membrane of the target cell. Instead of activating intracellular enzymes, however, they bind to receptor proteins located either in the cytosol or in the nucleus. These hormone receptors are proteins capable of regulating gene transcription, but they are typically present in an inactive form in unstimulated cells.

When a hormone binds, the receptor protein undergoes a large conformational change that activates the protein, allowing it to promote or inhibit the transcription of a selected set of genes. Each hormone binds a different receptor protein, and each receptor acts at a different set of regulatory sites in DNA. Because the hormones regulate different sets of genes, they evoke a variety of physiological responses. Steroid hormone receptors play an essential role in human physiology, as illustrated by the dramatic consequences of a lack of the receptor for testosterone in humans. The male sex hormone testosterone shapes the formation of the external genitalia and influences brain development in the fetus; at puberty, it triggers the development of male secondary sexual characteristics.

Some very rare individuals are genetically male (that is, they have both an X and a Y chromosome) but lack the testosterone receptor as a result of a mutation in the corresponding gene; thus they make the hormone, but their cells cannot respond to it. As a result, these individuals develop the appearance and behavior of females, which is the pathway of sexual and brain development that would occur if no male or female hormones were produced. This demonstrates the key role of the testosterone receptor in sexual development, and also shows that the receptor is required not just in one cell type to mediate one effect of testosterone, but in many cell types to help produce the whole range of features that distinguish men from women.

Classes of Cell-Surface Receptors

In contrast to NO and the steroid and thyroid hormones, the vast majority of signal molecules are too large or hydrophilic to cross the plasma membrane of the target cell. These proteins, peptides, and other bulky, water-soluble molecules bind to receptor proteins that span the plasma membrane (Figure 4). The transmembrane receptors detect a signal on the outside and relay the message, in a new form, across the membrane into the interior of the cell.

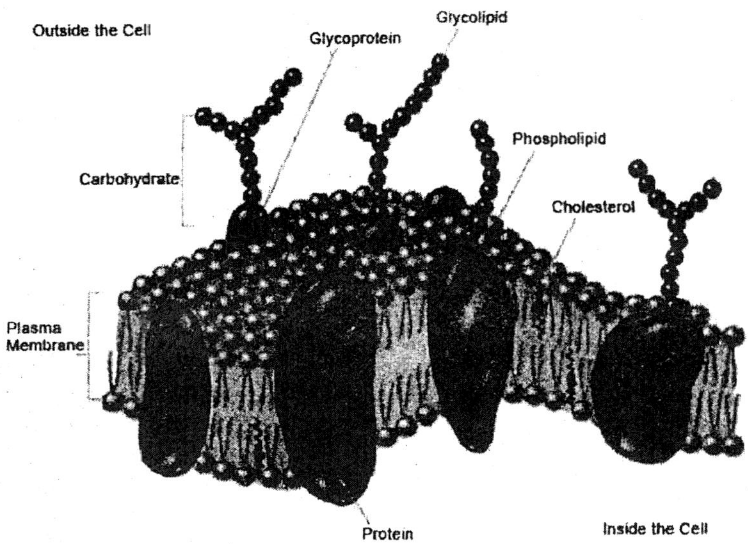

Figure 4. Plasma Membrane

Most cell-surface receptor proteins belong to one of three large families: ion-channel-linked receptors, G-protein-linked receptors, or enzyme-linked receptors. These families differ in the nature of the intra-cellular signal that they generate when the extracellular signal molecule binds to them. For ion-channel-linked receptors, the resulting signal is a flow of ions across the membrane, which produces an electrical current. G-protein-linked receptors activate a class of membrane-bound protein (a trimeric GTP-binding protein or G protein), which is then released to migrate in the plane of the plasma membrane, initiating a cascade of other effects.

And enzyme-linked receptors, when activated, act as enzymes or are associated with enzymes inside the cell.

Switching on this enzymatic activity then generates a host of additional signals, including small molecules that are released into the cytosol. The number of different types of receptors in these three classes is even greater than the number of extracellular signals that act on them, because for many extracellular signal molecules there is more than one type of receptor.

The neurotransmitter acetylcholine, for example, acts on skeletal muscle cells via an ion-channel-linked receptor, whereas in heart muscle cells it acts through a G-protein-linked receptor. These two types of receptors generate different intracellular signals, and thus enable the two types of muscle cells to react to acetylcholine in different ways, increasing contraction in skeletal muscle and decreasing the frequency of contractions in heart.

The multitude of different cell-surface receptors that the body requires for its own signalling purposes are also targets for many foreign substances that interfere with our physiology and sensations, from heroin and nicotine to tranquilizers and chili peppers. These substances either mimic the natural ligand for a receptor, occupying the normal ligand-binding site, or bind to the receptor at some other site, blocking or overstimulating the receptor's natural activity. Many drugs and poisons act in this way, and a large part of the pharmaceutical industry is devoted to the search for substances that will exert a precisely defined effect by binding to a specific type of cell-surface receptor.

Ion-channel-linked Receptors

Of all the cell-surface receptor types, ion-channel-linked receptors (also known as transmitter-gated ion channels) function in the simplest and most direct way. These receptors are responsible for the rapid transmission of signals across synapses in the nervous system. They transduce a chemical signal, in the form of a pulse of neurotransmitter delivered to the outside of the target cell, directly into an electrical signal, in the form of a change in voltage across the target cell's plasma membrane. When the neurotransmitter binds, this type of receptor alters its conformation so as to open or close a channel for the flow of specific types of ions-such as Na^+, K^+, Ca^{2+}, or Cl^-—across the membrane.

Driven by an electrochemical gradient, the ions rush into or out of the cell, creating a change in the membrane potential within a millisecond or so. This change in potential may trigger a nerve impulse, or alter the ability of other signals to do so. The opening of Ca^{2+} channels has special effects, as changes in the intracellular Ca^{2+} concentration can profoundly alter the activities of many enzymes. Whereas ion-channel-linked receptors are a specialty of the nervous system and of other electrically excitable cells such as muscle, Gprotein -linked receptors and enzyme-linked receptors are used by practically every cell type of the body.

Intracellular Signalling Proteins Act as Molecular Switches

Signals received via G-protein-linked or enzyme-linked receptors are transmitted to elaborate relay systems formed from cascades of intracellular signalling molecules. Apart from a few small molecules (such as cyclic GMP, cyclic AMP, and Ca^{2+}), these intracellular signalling molecules are proteins. Some serve as chemical transducers: in response to one type of chemical signal they generate another. Others serve as messengers, receiving a signal in one part of the cell and moving to another to exert an effect; and so on.

Most of the key intracellular signalling proteins behave as molecular switches: receipt of a signal switches them from an inactive to an active state. Once activated, these proteins can turn on other proteins in the pathway. They then persist in an active state until some other process switches them off again. The importance of the switching-off process is often underappreciated. If a signalling pathway is to recover after transmitting a signal and make itself ready to transmit another, every molecular switch must be reset to its original, unstimulated state. Thus, at every step, for every activation mechanism there has to be an inactivation mechanism. The two are equally important for the function of the system.

Proteins that act as molecular switches mostly fall into one of two main classes. The first and by far the largest class consists of proteins whose activity is turned on or off by phosphorylation.. For these, the switch is thrown in one direction by a protein kinase, which tacks a phosphate onto the switch protein, and in the other direction by a protein phosphatase, which plucks the phosphate off the switch protein. Many of the switch proteins controlled by phosphorylation are themselves protein kinases, and these are often organized into phosphorylation cascades: one protein kinase, activated by phosphorylation, phosphorylates the next protein kinase in the sequence, and so on, transmitting the signal onward and, in the process, amplifying it, distributing it, and modulating it.

The other main class of switch proteins involved in signalling consists of GTP-binding proteins. These switch between an active and an inactive state according to whether they have GTP or GDP bound to them. The mechanisms that control the switch on and the switch off will be described in the next section. GTP-binding proteins are important in several signalling pathways. One type, the G proteins, have a central role in signalling via G-protein-linked receptors.

G-PROTEIN-LINKED RECEPTORS

G-protein-linked receptors form the largest family of cell-surface receptors, with hundreds of members already identified in mammalian cells. They mediate responses to an enormous diversity of extracellular signal molecules, including hormones, local mediators, and neurotransmitters. These signal molecules are as varied in structure as they are in function: they can be proteins, small peptides, or derivatives of amino acids or fatty acids, and for each one of them there is a different receptor or set of receptors. Despite the diversity of the signal molecules that bind to them, all G-protein-linked receptors that have been analyzed possess a similar structure: each is made of a single polypeptide chain that threads back and forth across the lipid bilayer seven times.

This superfamily of seven-pass transmembrane receptor proteins includes rhodopsin (the light-activated photoreceptor protein in the vertebrate eye), the olfactory (smell) receptors in the vertebrate nose, and the receptors that participate in the mating rituals of single-celled yeasts. Evolutionarily speaking, G-protein-linked receptors are ancient: even bacteria possess structurally similar membrane proteins-such as the bacteriorhodopsin that functions as a light-driven H^+ pump. Although they resemble eucaryotic G-protein-linked receptors, these bacterial receptors do not act through G proteins; instead they are coupled to different signal transduction systems.

Stimulation of G-protein-linked Receptors

When an extracellular signalling molecule binds to a seven-pass transmembrane receptor, the receptor protein undergoes a conformational change that enables the receptor to activate a G protein located on the underside of the plasma membrane. To explain how this activation leads to the transmission of a signal, we must first consider how G proteins are constructed and how they function. There are several varieties of G proteins. All of these G proteins, however, have a similar general structure and operate in a similar way. They are composed of three protein subunits-α, β, and γ-two of which are tethered to the plasma membrane by short lipid tails. In the unstimulated state, the a subunit has GDP bound to it, and the G protein is idle.

When an extracellular ligand binds to its receptor, the altered receptor activates a G protein by causing the a subunit to lose some of its affinity for GDP, which it exchanges for a molecule of GTP. This activation breaks up the G protein subunits: the "switched-on" a subunit, clutching its GTP, detaches from the bg complex, giving rise to two separate molecules that now roam independently along the plasma membrane.

The two activated parts of a G protein-the a subunit and the βγ complex-can both interact directly with target proteins located in the plasma membrane, which in turn may relay the signal to yet other destinations. The longer these target proteins have an a or a bg subunit bound to them, the stronger and more prolonged the relayed signal will be.

The amount of time that the α and βγ subunits remain dissociated—and hence available to relay signals-is limited by the behavior of the a subunit. The α subunit has an intrinsic GTP-hydrolyzing (GTPase) activity, and it eventually hydrolyzes its bound GTP back to GDP; the α subunit then reassociates with a βγ complex and the signal is shut off. This reunion generally occurs within seconds after the G protein has been activated. The reconstituted G protein is now ready to be reactivated by another activated receptor.

Again, this system demonstrates a general principle of cell signalling: the mechanisms that shut a signal off are as important as the mechanisms that turn it on. They offer as many opportunities for control, and as many dangers of mishap. Take cholera, for example. The disease is caused by a bacterium that multiplies in the intestine, where it produces a protein called cholera toxin. This protein enters the cells that line the intestine and modifies the a subunit of a G protein (called G_s, because it stimulates the enzyme adenylyl cyclase, discussed later) in such a way that it can no longer hydrolyze its bound GTP. The altered a subunit thus remains in the active state indefinitely, continuously transmitting a signal to its target proteins. In intestinal cells, this causes a prolonged and excessive outflow of Cl^- and water into the gut, resulting in catastrophic diarrhea and dehydration. The condition often leads to death unless urgent steps are taken to replace the lost water and ions.

A similar situation occurs in whooping cough (pertussis), a common respiratory infection against which infants are now routinely vaccinated. In this case, the disease-causing bacterium colonizes the lung, where it produces a protein called pertussis toxin. This protein alters the a subunit of a different type of G protein (called G_i, because it inhibits adenylyl cyclase). In this case, however, modification by the toxin disables the G protein by locking it into its inactive GDP-bound state. Knocking out G_i, like activating G_s, results in the generation of a prolonged, inappropriate signal. Oddly, although the biochemical effects of cholera and pertussis toxins are known in detail, it is not clear how the bacteria benefit from their actions. In any case, what cholera and pertussis toxins do show us is that, like a car accelerating out of control, intracellular signalling pathways can be rendered dangerously overactive either by gluing down the molecular gas pedal or by cutting the molecular brakes.

G Proteins Regulate Ion Channels

The target proteins for G-protein subunits are either ion channels or membrane-bound enzymes. Different targets are affected by different types of G proteins (of which about 20 have so far been discovered in mammalian cells), and these various G proteins are themselves activated by different classes of cell-surface receptors. In this way, binding of an extracellular signal molecule to a G-protein-linked receptor leads to effects on a particular subset of the possible target proteins, eliciting a response that is appropriate for that signal and that type of cell. An example of G-protein regulation is ion channels. The heartbeat in animals is controlled by two sets of nerve fibers: one set speeds the heart up, the other slows it down. The nerves that signal a slowdown in heartbeat do so by releasing acetylcholine, which binds to a G-protein-linked receptor on the surface of the heart muscle cells.

When acetylcholine binds to this receptor, a G protein (G_i) is activated-dissociating into an a subunit and a bg complex. In this particular example, the bg complex is the active signalling component: it binds to the intracellular face of a K^+ channel in the heart muscle cell plasma membrane, forcing the ion channel into an open conformation and allowing K^+ to flow out of the cell. This alters the electrical properties of the heart muscle cell, inhibiting its activity. The signal is shut down-and the K^+ channel recloses-when the a subunit inactivates itself by hydrolyzing its bound GTP and reassociates with the bg complex to form an inactive G protein.

Activated Membrane-bound Enzymes

The interactions of G proteins with ion channels cause an immediate change in the state and behavior of the cell. Their interactions with enzyme targets have more complex consequences, leading to the production of additional intracellular signalling molecules. The most frequent target enzymes for G proteins are adenylyl cyclase, the enzyme responsible for production of the small intracellular signalling molecule cyclic AMP, and phospholipase C, the enzyme responsible for production of the small intracellular signalling molecules inositol trisphosphate and diacylglycerol. These two enzymes are activated by different types of G proteins, so that cells are able to couple the production of the small intracellular signalling molecules to different extracellular signals.

The coupling may be either stimulatory or inhibitory. G proteins that stimulate enzyme activity. The small intracellular signalling molecules generated in these cascades are often called second messengers (the "first messengers" being the extracellular signals); they are produced in large numbers when a membrane-bound enzyme-such as adenylyl cyclase or phospholipase C-is activated, and they rapidly diffuse away from their source, spreading the signal throughout the cell.

Different second-messenger molecules, of course, produce different cellular responses. The consequences of an increase in the intracellular concentration of cyclic AMP. This will take us along one of the main types of signalling pathways that lead from the activation of G-protein-linked receptors.

Cyclic AMP Pathway

Many extracellular signals acting via G-protein-linked receptors affect the activity of adenylyl cyclase and thus alter the concentration of the messenger molecule cyclic AMP inside the cell. Most commonly, the activated G-protein a subunit switches on the adenylyl cyclase, causing a dramatic and sudden increase in the synthesis of cyclic AMP from ATP (which is always present in the cell). Because it stimulates the cyclase, this G protein is called Gs. To eliminate the signal, a second enzyme, called cyclic AMP phosphodiesterase, rapidly converts cyclic AMP to ordinary AMP. One way that caffeine acts as a stimulant is by inhibiting this enzyme in the nervous system, blocking cyclic AMP degradation and keeping the concentration of this second-messenger molecule high.

Cyclic AMP phosphodiesterase is continuously active inside the cell. Because it breaks cyclic AMP down so quickly, the concentrations of this second messenger can change rapidly in response to extracellular signals, rising or falling tenfold in a matter of seconds. Cyclic AMP is a water-soluble molecule, so it can carry its signal throughout the cell, traveling from the site on the membrane where it is synthesized to interact with proteins located in the cytosol, the nucleus, or other organelles.

Many cell responses are mediated by cyclic AMP; a few are listed in Table 1. As the table shows, different target cells respond very differently to extracellular signals that change intracellular cyclic AMP levels. In many types of animal cells, stimulating cyclic AMP production boosts the rate of consumption of metabolic fuel. When we are frightened or excited, for example, the adrenal gland releases the hormone adrenaline, which circulates in the bloodstream and binds to a class of G-protein-linked receptors (adrenergic receptors) that are present on many types of cells.

Table 1. Some Hormone-induced Cell Responses Mediated by Cyclic AMP

Extracellular Signal Molecule	Target Tissue	Major Response
Adrenaline	heart	increase in heart rate and force of contraction
Adrenaline	muscle	glycogen breakdown
ACTH, glucagon	fat	fat breakdown
ACTH adrenal	gland	cortisol secretion

The consequences vary from one cell type to another, but all the cell responses help prepare the body for sudden action. In skeletal muscle, for example, adrenaline triggers a rise in the intracellular level of cyclic AMP, which causes the breakdown of glycogen (the polymerized storage form of glucose). This glycogen breakdown makes more glucose available as fuel for anticipated muscular activity. Adrenaline also acts on fat cells, stimulating the breakdown of triacylglyceride (the storage form of fat) to fatty acids-an immediately usable form of cell fuel, which can also be exported to other cells.

Cyclic AMP exerts these various effects mainly by activating the enzyme cyclic-AMP-dependent protein kinase (PKA). This enzyme is normally held inactive in a complex with another protein. The binding of cyclic AMP forces a conformational change that unleashes the active kinase. Activated PKA then catalyzes the phosphorylation of particular serines or threonines on certain intracellular proteins, thus altering their activity. In different cell types, different sets of target proteins are available to be phosphorylated, which explains why the effects of cyclic AMP vary with the target cell.

In some cases the effects of activating a cyclic AMP cascade are rapid; in others the effects are slow. In skeletal muscle cells, for example, activated PKA phosphorylates enzymes involved in glycogen metabolism, triggering the mechanism that breaks down glycogen to glucose. This response occurs within seconds. At the other extreme, some cyclic AMP responses take minutes or hours to develop. Included in this slow class are responses that involve changes in gene expression, an important means of regulating cell behavior.

Thus in some cells, the PKA phosphorylates gene regulatory proteins that then activate the transcription of selected genes, a process that requires minutes or hours. In endocrine cells in the hypothalamus, for example, a rise in the amount of intracellular cyclic AMP stimulates the production and secretion of a peptide hormone called somatostatin. Increases in cyclic AMP concentrations in neurons, by contrast, control the production of proteins involved in long-term memory.

Inositol Phospholipid Pathway Triggers

Some G-protein-linked receptors exert their effects via a type of G protein that activates the membrane-bound enzyme phospholipase C instead of adenylyl cyclase. Once activated, phospholipase C propagates its signal by cleaving a lipid molecule that is a component of the cell membrane. The molecule is an inositol phospholipid (a phospholipid that has the sugar inositol attached to its head) that is present in small quantities in the inner half of the plasma membrane lipid bilayer. Because of the involvement of this phospholipid, the signalling pathway that begins with the activation of phospholipase C is often known as the inositol phospholipid pathway. This signalling cascade occurs in almost all eucaryotic cells and affects a host of different target proteins.

The cascade works in the following way. When phospholipase C chops the sugar-phosphate head off the inositol phospholipid, it generates two small messenger molecules-inositol 1,4,5-trisphosphate (IP3) and diacylglycerol (DAG). IP3, a hydrophilic sugar phosphate, diffuses into the cytosol, while the lipid DAG remains embedded in the plasma membrane. Both molecules play a crucial part in signalling inside the cell, and we will consider them in turn. The IP3 released into the cytosol will eventually encounter the endoplasmic reticulum; there it binds to and opens Ca^{2+} channels that are embedded in the endoplasmic reticulum membrane. Ca^{2+} stored inside the endoplasmic reticulum rushes out into the cytosol through these open channels, causing a sharp rise in the cytosolic concentration of free Ca^{2+}, which is normally kept very low.

Together with Ca^{2+}, diacylglycerol helps recruit and activate a protein kinase which translocates from the cytosol to the plasma membrane, This enzyme is called protein kinase C (PKC) because it also needs to bind Ca^{2+} to become active. Once activated, PKC phosphorylates a set of intracellular proteins that varies depending on the cell type. PKC operates on the same principle as PKA, although most of its target proteins are different.

Ca^{2+} has such an important and widespread role as an intracellular messenger that we must digress to consider its functions more generally. A surge in the cytosolic concentration of free Ca^{2+} is triggered by many different signals, not only those that act through Gprotein linked receptors. When a sperm fertilizes an egg cell, for example, Ca^{2+} channels open, and the resulting rise in cytosolic Ca^{2+} triggers the start of embryonic development; for skeletal muscle cells, a signal from a neighboring nerve triggers a rise in cytosolic Ca^{2+} that initiates contraction; and in many secretory cells, including nerve cells, Ca^{2+} triggers secretion.

Ca^{2+} stimulates all these responses by binding to and influencing the activity of Ca^{2+}-sensitive proteins. The concentration of free Ca^{2+} in the cytosol of an unstimulated cell is extremely low (10^{-7} M) compared with its concentration in the extracellular fluid and in the endoplasmic reticulum. These differences are maintained by membrane pumps that actively pump Ca^{2+} out of the cytosol-either into the endoplasmic reticulum or across the plasma membrane and out of the cell. As a result, a steep electrochemical gradient of Ca^{2+} exists across the endoplasmic reticulum membrane and across the plasma membrane.

When a signal transiently opens Ca^{2+} channels in either of these membranes, Ca^{2+} rushes into the cytosol down its electrochemical gradient, triggering changes in Ca^{2+}-responsive proteins in the cytosol. The effects of Ca^{2+} in the cytosol are largely indirect: they are mediated through the interaction of Ca^{2+} with various transducer proteins, known collectively as Ca^{2+}-binding proteins. The most widespread and common of these is the Ca^{2+}-responsive protein calmodulin. Calmodulin is present in the cytosol of all eucaryotic cells that have been examined, including those of plants, fungi, and protozoa.

When calmodulin binds to Ca^{2+}, the protein undergoes a conformational change that enables it to wrap around a wide range of target proteins in the cell, altering their activity. One particularly important class of targets for calmodulin includes the Ca^{2+}/calmodulindependent protein kinases (CaM-kinases). When these kinases are activated by binding to calmodulin complexed with Ca^{2+}, they influence other processes in the cell by phosphorylating selected proteins. In the mammalian brain, for example, a neuron-specific CaM-kinase is abundant at synapses, where it is thought to play a part in learning and memory.

Intracellular Signalling Cascades

The steps in the signalling cascades associated with G-protein-linked receptors take a long time to describe, but they often take only seconds to execute. Consider how quickly a thrill can make your heart beat faster (when adrenaline stimulates the G-protein-linked receptors in your heart muscle cells, accelerating your heartbeat), or how fast the smell of food can make you salivate (through the G-protein-linked receptors for odors in your nose and the G-protein-linked receptors for acetylcholine in salivary cells that stimulate secretion).

Among the fastest of all responses mediated by a G-protein-linked receptor, however, is the response of the eye to bright light: it takes only 20 msec for the most quickly responding photoreceptor cells of the retina (the cone photoreceptors) to produce their electrical response to a sudden flash of light. This speed is achieved in spite of the necessity to relay the signal over several steps of an intracellular signalling cascade. But photoreceptors also provide a beautiful illustration of the positive advantages of signalling cascades: in particular, such cascades allow spectacular amplification of the incoming signal and allow cells to adapt so as to be able to detect signals of widely varying intensity.

The quantitative details have been most thoroughly analyzed for the rod photoreceptor cells in the eye. Here, light is harvested by rhodopsin, a Gprotein -linked light receptor. Light-activated rhodopsin activates a G protein called transducin. The activated a subunit of transducin then activates an intracellular signalling cascade that causes Na+ channels to close in the plasma membrane of the photoreceptor cell. This produces a change in the voltage across the cell membrane, with the ultimate consequence that a nerve impulse is sent to the brain. The signal is repeatedly amplified as it is relayed along this pathway.

The adaptation depends on negative feedback: an intense response in the photoreceptor cell generates an intracellular signal (a change in Ca^{2+} concentration) that inhibits the enzymes responsible for signal amplification. Adaptation also occurs in signalling pathways that respond to chemical signals; again, it allows cells to remain sensitive to changes of signal intensity over a wide range of background levels of stimulation. Adaptation, in other words, allows a cell to respond to both messages that are whispered and those that are shouted. In addition to vision, taste and smell also depend on Gprotein linked receptors. It seems likely that this mechanism of signal reception, invented early in the evolution of the eucaryotes, has its origins in the basic and universal need of cells to sense and respond to their environment.

REFERENCES

Clarke M.B. and V. Sperandio, "Events at the host-microbial interface of the gastrointestinal tract III", Cell-to-cell signalling among microbial flora, host, and pathogens: there is a whole lot of talking going on" in *American journal of physiology, Gastrointestinal and liver physiology*, 2005.

Mohamed O. A., M. Jonnaert, C. Labelle-Dumais, K. Kuroda, H. J. Clarke and D. Dufort, "Uterine Wnt/beta-catenin signalling is required for implantation", in *Proceedings of the National Academy of Sciences of the United States of America Volume*, 2005.

5

Cell Transcription

Transcription is the synthesis of RNA under the direction of DNA. Both nucleic acid sequences use the same language, and the information is simply transcribed, or copied, from one molecule to the other. DNA sequence is enzymatically copied by RNA polymerase to produce a complementary nucleotide RNA strand, called messenger RNA (mRNA), because it carries a genetic message from the DNA to the protein-synthesizing machinery of the cell. One significant difference between RNA and DNA sequence is the presence of U, or uracil in RNA instead of the T, or thymine of DNA. In the case of protein-encoding DNA, transcription is the first step that usually leads to the expression of the genes, by the production of the mRNA intermediate, which is a faithful transcript of the gene's protein-building instruction.

The stretch of DNA that is transcribed into an RNA molecule is called a *transcription unit*. A transcription unit that is translated into protein contains sequences that direct and regulate protein synthesis in addition to coding the sequence that is translated into protein. The regulatory sequence that is before, or 5', of the coding sequence is called 5' untranslated (5'UTR) sequence, and sequence found following, or 3', of the coding sequence is called 3' untranslated (3'UTR) sequence. Transcription has some proofreading mechanisms, but they are fewer and less effective than the controls for copying DNA; therefore, transcription has a lower copying fidelity than DNA replication.

As in DNA replication, transcription proceeds in the 5' → 3' direction. Only one of the two DNA strands is transcribed. This strand is called the template strand, because it provides the template for ordering the sequence of nucleotides in an RNA transcript. The DNA template strand is read 3' → 5' by RNA polymerase and the new RNA strand is synthesized in the 5'→ 3' direction. RNA polymerase binds to the 5' end of a gene (promoter) on the DNA template strand and travels toward the 3' end.

Except for the fact that thymines in DNA are represented as uracils in RNA, the newly synthesized RNA strand will have the same sequence as the coding (non-template) strand of the DNA. For this reason, scientists usually refer to the DNA coding strand that has the same sequence as the resulting RNA when referring to the directionality of genes on DNA, not the template strand.

PROKARYOTIC TRANSCRIPTION

Prokaryotic transcription is the process in which messenger RNA transcripts of genetic material in prokaryotes are produced, to be translated for the production of proteins. Prokaryotic transcription occurs in the cytoplasm alongside translation. Unlike in eukaryotes, prokaryotic transcription and translation can occur simultaneously. This is impossible in eukaryotes, where transcription occurs in a membrane-bound nucleus while translation occurs outside the nucleus in the cytoplasm. In prokaryotes genetic material is not enclosed in a membrane-enclosed nucleus and has access to ribosomes in the cytoplasm.

Initiation

The following steps occur, in order, for transcription initiation:

- RNA polymerase (RNAP) binds to one of several specificity factors, s, to form a holoenzyme. In this form, it can recognize and bind to specific promoter regions in the DNA. At this stage, the DNA is double-stranded ("closed"). This holoenzyme/wound-DNA structure is referred to as the *closed complex.*
- The DNA is unwound and becomes single-stranded ("open") in the vicinity of the initiation site (defined as +1). This holoenzyme/unwound-DNA structure is called the *open complex.*
- The RNA polymerase transcribes the DNA, but produces about 10 abortive (short, non-productive) transcripts which are unable to leave the RNA polymerase because the exit channel is blocked by the s-factor.
- The s-factor eventually dissociates from the holoenzyme, and elongation proceeds.

Promoters can differ in "strength"; that is, how actively they promote transcription of their adjacent DNA sequence. Promoter strength is in many (but not all) cases, a matter of how tightly RNA polymerase and its associated accessory proteins bind to their respective DNA sequences. The more similar the sequences are to a consensus sequence, the stronger the binding is. Additional transcription regulation comes from transcription factors that can affect the stability of the holoenzyme structure at initiation.

Most transcripts originate using adenosine-5'-triphosphate (ATP) and, to a lesser extent, guanosine-5'-triphosphate (GTP) (purine nucleoside triphosphates) at the +1 site. Uridine-5'-triphosphate (UTP) and cytidine-5'-triphosphate (CTP) (pyrimidine nucleoside triphosphates) are disfavoured at the initiation site.

Termination

Two termination mechanisms are well known:

— Intrinsic termination (also called Rho-independent transcription termination) involves terminator sequences within the RNA that signal the RNA polymerase to stop. The terminator sequence is usually a palindromic sequence that forms a stem-loop hairpin structure that leads to the dissociation of the RNAP from the DNA template.

— Rho-dependent termination uses a termination factor called *r* factor(rho factor) which is a protein to stop RNA synthesis at specific sites. This protein binds at a rho ultilisation site on the nascent RNA strand and runs along the mRNA towards the RNAP. A stem loop structure upstream of the terminator region pauses the RNAP, when *r*-factor reaches the RNAP, it causes RNAP to dissociate from the DNA, terminating transcription.

Other termination mechanisms include where RNAP comes across a region with repetitious thymidine residues in the DNA template, or where a GC-rich inverted repeat followed by 4 A residues. The inverted repeat forms a stable stem loop structure in the RNA, which causes the RNA to dissociate from the DNA template.

The -35 region and the -10 ("Pribnow box") region comprise the basic prokaryotic *promoter*, and [T] stands for the *terminator*. The DNA on the template strand between the +1 site and the terminator is transcribed into RNA, which is then translated into protein..

EUKARYOTIC TRANSCRIPTION

Eukaryotic transcription is more complex than prokaryotic transcription. For instance, in eukaryotes the genetic material (DNA), and therefore transcription, is primarily localized to the nucleus, where it is separated from the cytoplasm (in which translation occurs) by the nuclear membrane. DNA is also present in mitochondria in the cytoplasm and mitochondria utilize a specialized RNA polymerase for transcription. This allows for the temporal regulation of gene expression through the sequestration of the RNA in the nucleus, and allows for selective transport of RNAs to the cytoplasm, where the ribosomes reside.

The basal eukaryotic transcription complex includes the RNA polymerase and additional proteins that are necessary for correct initiation and elongation.

Initiation

Among eukaryotes that regulate the transcription of individual genes, the *core promoter* of protein-encoding gene contain binding sites for the basal transcription complex and RNA polymerase II, and is normally within about 50 bases upstream of the transcription initiation site. Further transcriptional regulation is provided by *upstream control elements* (UCEs), usually present within about 200 bases upstream of the initiation site. The core promoter for Pol II sometimes contains a TATA box, the highly conserved DNA recognition sequence for the TATA box binding protein, TBP, whose binding initiates transcription complex assembly at the promoter.

Some genes also have enhancer elements that can be thousands of bases upstream or downstream of the transcription initiation site. Combinations of these upstream control elements and enhancers regulate and amplify the formation of the basal transcription complex.

Eukaryotic transcription is carried out by three separate RNA polymerases. RNAPI transcribes the precursors of several ribosomal RNAs. RNAPIII transcribes an additional ribosomal RNA, 5S RNA, and all tRNAs. RNAPII transcribes the mRNA encoding proteins and the RNA of the small ribonucleoprotein particles. In vitro, transcription of eukaryotic DNA by RNAPII alone is initiated at random sites. Initiation at the proper site requires the presence of accessory proteins. These include as many as seven general transcription factors (GTFs).

Transcription factors (TFs) are numbered to identify the RNAP with which they interact and a letter to distinguish one from the other (e.g., TFIIA). A highly ordered assembly of GTFs, RNAPII and promoter DNA of the expressed gene are required for initiation of transcription of mRNA. In addition to transcription, the production of mature mRNA requires several other processing reactions. The primary transcript, the RNA encoded by the DNA of an entire gene, is spliced so that parts of the molecule are removed.

At the 5' end of each molecule a 7-methyl guanosine residue cap is attached to the RNA by triphosphate linkage, a reaction catalyzed by a specific guanylyl transferase enzyme. The 3' end of the mRNA is cleaved downstream from the AAUAAA sequence and then a long run of A [a poly(A) tail] is attached to the 3' ends. The maturation processes depend on the carboxy-terminal of the polymerase. Apparently RNAPII forms a complex with processing factors active in the maturing processes including the capping of the mRNA.

Proteins and DNA Interactions

There are many ways in which binding can be detected. Electrophoretic migration in a polyacrylamide gel is a function of molecular size, where the gel acts as a sieve: the larger the molecule, the slower the migration. Therefore, DNA fragments would migrate more slowly if bound to a protein. When DNA is exposed to a disruptive procedure, such as a special chemical treatment or exposure to endonuclease, the region of the DNA which binds to a protein is protected from breakdown. The protected piece will not be fragmented and will not appear in the gel (which excludes large pieces) after electrophoresis.

In practice, the portions of the DNA exposed to the treatment in the presence of protein, is compared to that which was treated in the absence of protein. The DNA fragments will appear as bands. When the corresponding sectors of the gels are compared, an empty space will appear where the DNA was protected by the protein binding. The various procedures have identified binding sites in the DNA and the construction of maps showing their location.

FACTORS OF TRANSCRIPTION

In vitro experiments have revealed that in addition to RNAPII as many as seven general transcription factors (GTFs) are required for transcription of mRNA. TFIID, which is needed for initiation of transcription, binds the TATA box of many promoters. It is a complex which includes a TATA binding protein (TBP), as well as eight other subunits known as TBP-associated factors (TAFs). The assembly of the transcription machinery proceeds in an ordered process. TFIID, TBP and associated factors first form a complex. Other GTFs and RNAPII are then incorporated, forming a pre-initiation complex.

In the presence of all four ribonucleotide triphosphates, RNAPII, in the pre-initiation complex, can then initiate transcription at the start site. The number of GTFs required may differ from promoter to promoter site, and there is some evidence that the exact complement needed depends on the structure of the DNA. The GTFs also mediate the promoter melting (DNA unwinding) and the transition between initiation and elongation of the RNA.

TFIID has the DNA-specific binding capacity to recognize promoters and to mediate regulatory signals from upstream activators. Part of the complex, $TAF_{II}250$, possesses two tandem bromodomain modules that bind selectively to acetylated histone H4 peptides. Acetylation-deacetylation is thought to be part of the mechanisms of transcriptional regulation, where acetylation is thought to make nucleosomal DNA more accessible to transcription. Bromodomains are modules of approximately 110-amino-acid present in many chromatin-associated proteins.

The crystal structure of the double bromodomain shows two four-helix bundles with a highly asymmetric surface charge distribution. Each bundle contains a structure ideally suited for recognition of diacetylated histone H4 tails. The activation of the appropriate gene could then proceed as follows. First a histone acetyltransferase (HAT) containing activator is recruited. After recruitment, the amino terminal histone tails of the nucleosomes are acetylated. The acetylated tails then recruit TFIID via the bromodomains of $TAF_{II}250$. The role of $TAF_{II}250$ may be more immediate since this complex has been shown to have HAT-activity. TFs generally contain two domains which operate independently.

One domain binds DNA, the other activates transcription (the TAD domain). The independence of the two functions has been shown by producing hybrid proteins, where the DNA binding site of one protein is combined with the activation site of another. The resulting protein is fully active but now the biological signal activates the wrong gene.

The mechanisms by which transcription factors increase transcription are still unknown; they are presumed to function in stabilizing the binding between a transcription complex containing RNAP and the DNA. The RNA polymerase, TBP and various transcription factors constitute the minimal, basic machinery for transcription. In addition, activator complexes attaching upstream of the TATA box, speed up the transcription either by direct binding or via a mediator. TFIID is thought to be recruited to the promoter by activators. TFIID binds the TATA nucleotide sequence. This is because the TFIID complex includes TBP and TAFs which bridge DNA-specific-activators and the basic transcriptional machinery.

Although transcription factors are most frequently involved in activation of gene expression, they also can serve as repressors. Others can act either as activators or repressors depending on individual conditions. The TFs act by binding either coactivators or corepressor proteins. An example of this double function is provided by studies of the transcriptional factor Pt-1. Pt-1 is required for the development of the pituitary gland and is involved in the regulation of the genes coding for growth hormone, prolactin and thyrotropin. The factor binds both the growth hormone and the pituitary promoters. In lactotropes (anterior pituitary prolactin-secreting cells), Pit-1 activates the pituitary promoter and in addition, represses the growth hormone promoter by allowing the binding of the corepressor N-CoR.

Generally, transcription factors are rapidly degraded by the ubiquitin-proteasome system. The effectiveness of a transcription factor correlates highly with its rate of degradation, suggesting a link between the two activities. A connection between the two processes is farther supported by the finding that mutations in the TAD domain of the transcription factor produce parallel changes in transcription activation and degradation. VP16 is an activator of the transcription of five virally encoded genes of herpes simplex virus type 1. A transcription factor containing the VP16 TAD signals ubiquitination through the met30 ubiquitin-ligase. In addition met30 is needed for V16 TAD to activate transcription. In the absence of a met30, a construct of VP16 fused to ubiquitin activates transcription, demonstrating that ubiquitination is needed for transcriptional activity.

Several cofactors have been implicated in the regulation of the general transcriptional machinery, such as the universal stimulatory activity (USA) found in extracts and which contains negative and positive cofactors. Some of these interact directly with TBP. A set of RNAP II coactivator complexes, Mediators, are sequence specific activators that function in recruiting RNAP II to the promoter region of specific target genes. Some activators may act through Mediator complexes. A number of coactivators have been found to be large multifunctional complexes. They function as adaptors mediating activator recruitment of the transcriptional apparatus. Some are chromatin-remodeling or - modifying enzymes.

Gene expression has been found to be regulated by remodeling of chromatin, which may take place by acetylation, phosphorylation or methylation of histones, or chromatin remodeling enzymes which depend on the hydrolysis of ATP. Surprisingly, covalent modifications of the coactivators, including acetylation, phosphorylation and methylation, have also been found to act in transcriptional regulation. Coactivators may activate a variety of transcription factors or to have very specific effects. The modification of cofactors has been found to be carried out at least in part by the same enzymes involved with the covalent modification of histones.

Eukaryotic Promoters

In prokaryotes, the promoter sequence, at the beginning of the operon, is the site at which transcription is initiated by binding RNAP. Similarly, in eukaryotes, RNAP II promoters are needed for the accurate and efficient transcription of mRNA. As knowledge of the base sequence of eukaryotic DNA expanded, it was natural to look for recurring base sequence motifs in the DNA of the region upstream from the structural genes. Many such regions with special recurring motifs were identified. But how many of these are actually promoters? How can promoter function be demonstrated? Generally, a suspected promoter region has been studied by mutagenesis of the DNA piece and introduction into a living cell. This is most frequently done by attaching the DNA to a plasmid vector.

A plasmid is a piece of DNA that is capable of dividing independently from the chromosomes. The transfer of DNA into living cells using vectors (frequently of viral origin) is known as transfection. Most often a reporter gene is inserted downstream from the suspected sequence. The reporter gene is one that is easily recognizable by a simple assay, for example, an enzyme activity producing a color reaction. A favorite has been the *lacZ* gene of *E. coli,* which codes for the enzyme b-galactosidase and hydrolyses lactose to its component monosaccharides.

A block in the transcription of the reporter gene produced by a mutation in the suspected promoter region, is good evidence that it functions as a promoter. Typical eukaryotic promoters contain an AT region (the TATA box) and one or more sequences containing 8 to 12 base pair elements called the upstream promoter elements (UPE), such as CCAAT and GGCG. Apparently, the TATA box insures accurate initiation of transcription, whereas the UPE increases the rate of transcription. The TATA box is typically 25 to 30 base pairs upstream from the site of transcription, and the UPE are upstream from this.

We have seen that eukaryotic RNAPs require the presence of transcription complexes which include many accessory proteins and the promoter region of DNA. Promoters for RNAPIII are 50 nucleotide long within the 5S gene. The promoters for RNAPI and II, in contrast, are upstream 5' to the gene.

Enhancers

Animal viruses have been useful models of gene behavior. They can grow inside cells in culture and they borrow the entire array of the cell's machinery for their own duplication. Aside from a few proteins, they are basically constituted of a simple genetic DNA complement with as few as 5,000 base pairs. These DNA regions were called enhancers. Typically, enhancers contain 50 to 150 base pairs. They are composed of duplicated sequences in tandem. Apparently, the effect of enhancers does not depend on any single sequence motif. Synthetic oligonucleotides, stringing together repeats of oligonucleotides 18 to 20 base pairs in length, can functionally replace enhancers.

Enhancers play a role in differentiation and in responding to hormonal or environmental signals. The immunoglobulin genes expressed at high levels in B cells of the immune system involve enhancers. In contrast, in other cells not expressing the immunoglobulin genes, the enhancers serve to block transcription. Enhancers have been found to come in two different forms, either modular or non-modular. The modular enhancers can respond to different activators allowing for a variety of distinct patterns of gene expression in which each activator makes a distinct contributions to transcription.

Non-modular enhancers have an all-or-none function by acting as on-or-off switches where the "on" switch is turned on only when all activators are present. Certain activators may be present only in one cell type and during a particular developmental stage. Many activators can act on a single gene. In addition, the activators can be arranged in different combinations. The enhancer sequences of DNA contain multiple binding sites for activators. Before transcription may proceed, chromatin which was previously silenced must be remodeled to permit access to the transcriptional machinery.

Activators organize the recruitment of many proteins such as chromatinremodeling enzymes as well as proteins involved in the assembly of the core transcription machinery. The pattern generally recognized proceeds by a mechanism in which activators first recruit chromatinremodeling enzymes followed by the assembly of the preinitiation complex (PIC) which is constituted by RNA polymerase II and general transcription factors. Remodeling of chromatin may take place by the covalent modification of nucleosomal histones by acetylation, phosphorylation or methylation or by enzymes which change the chromatin structure directly and require the hydrolysis of ATP.

In many cases, gene transcription was found to require the assembly of transcription factors, activators and DNA complexes referred to as enhanceosomes. The IFN-b enhancer is a 65-base pair region upstream and adjacent to the promoter of the IFN-b gene. This enhancer is non-modular and initiates the transcription of the gene coding for IFN-b following viral infection. The assembly of the enhanceosome activates transcription by recruiting the histone acetyltransferase proteins, CREB binding protein (CBP) and p300/CBP-associated factors (PCAF)/GCN5 (proteins with histone acetyltransferase activity). In addition to modifying histones, these components acetylate, HMGI(Y), the architectural component required for enhanceosome assembly. The enhanceosome assembly occurs in well defined steps.

The stability of the complex is enhanced by acetylation at lysine-71 of HMGI(Y). The disassembly is automatic upon acetylation of HMGI(Y) (at lysine-65) which inhibits its ability to bind DNA. A similar role of enhanceosomes is suspected for several other mammalian genes see. However, a surprising amount of variation has been found in the details of the assembly of the necessary proteins. The most striking difference is that found for the activation of the human a_1-antitrypsin gene (a_1-*AT*) which involves the region upstream from the a_1-*AT* gene containing nucleosomes (e.g., the region upstream of the IFN-b gene involved in activation is free of nucleosomes). One activator protein and two general transcription factors are bound to the nucleosomal promoter region. When differentiation of the cell is induced, RNA polymerase II and general transcription factors are assembled in the PIC, in contrast to the general pattern, in absence of remodeling enzymes. However, a SWI/SNF and two HAT enzymes are recruited to the promoter region after PIC assembly and the promoter-associated nucleosomes are disrupted before transcription.

In reality, gene expression is controlled by a complex network which is not completely understood. A variety of *cis* elements such as the enhancers and silencers play a role. Silencers are genes which block the expression of other genes. Enhancers and silencers may reflect the structural organization of the chromatin. In addition to these elements, insulator elements have been identified which block interactions between cis-regulatory elements. These insulators are thought to have a role in separating active and and inactive genes that are expressed differentially. Insulator DNAs can work in a dominant fashion to block enhancerpromoter interactions over long distances. An additional set of regulatory elements has been identified that permit bypassing insulators. This element, the promoter targeting sequence (PTS), permits enhancers to overcome the blocking effects of insulators.

Structural Factors

Both single nucleosomes and high order chromosome structures are thought to inhibit transcription by a variety of mechanisms, such as steric hindrance of trans-acting factors binding or distortion of target binding sites. However, transcription can occur despite these inhibitions, suggesting that transcription factors can contend with either nucleosomes or folding.

In vitro, nucleosomes inhibit transcription in many cases. The inhibition can be reversed by a variety of mechanisms:

(1) transcription factors can bind newly replicated DNA before nucleosomes are formed,
(2) regulatory factors may bind to sites in the nucleosomes and disrupt them, and
(3) specialized disruptors may open the nucleosomes.

The concept that structural factors have a role in gene expression, namely that the disruption of the compacted form of chromatin is necessary for gene activation, is supported by experiments carried out in vivo. The lac repressor binding of direct repeats of the lac operator has allowed observing the organization of chromatin in vivo. GFP-repressor fusion proteins can be used at the light microscope level.

The method, combined with gene amplification has allowed studies of living cells. In these studies repeats of the lac operator were placed in vector constructs used for transfection. These sites, incorporated into the chromatin of the host cells, could then be labeled by the lac repressor-GFP chimeras that require detection of the fluorescence of this protein.

Using electron microscopy, the gene amplification regions were shown to correspond to 100 nm fibers. GFP-lac repressor constructs made it possible to carry out in vivo observations of the operator-tagged chromosomal DNA either when integrated by a vector or homologous recombination (yeast).Transcriptional activation and chromatin organization were studied using this methodology.

Lac operator repeats inserted in the chromosomes were targeted with a VP16-lac repressor fusion protein containing an VP16 acidic activation domain. This targeting resulted in increased transcription, localized histone hyperacetylation, and recruitment of at least three different histone acetyltransferases. These events are all associated with the activation of genes. The chromatin structure unfolded to form 25-40 mm fibers. The unfolding was propagated on a large-scale over hundreds of kilobase pairs. These changes occurred even when transcription was inhibited, suggesting that recruitment of the transcriptional machinery is responsible for the changes in chromatin structure.

In the experiments of Tsukamoto et al. a similar system was used with GFP. Genes in specifics location in the chromatin were activated and the actual production of a specific protein was demonstrated. Among these SWI/SNF, RSC, Brahma (BRM) and Imitation Switch (ISWI). HATs and HDACs by modifying the histone components of nucleosomes also affect their structure. Remodeling complexes as well as the complexes which modify histones covalently exhibit great specificity. Apparently, the specificity is provided by the association with other proteins, some of them DNA binding proteins

Mediator and Repression Complexes

In mammalian cells certain hormone receptors are activators. In the case of the thyroid hormone receptor (TR) the activation requires a mediator present in the TR-associated protein complex (TRAP). Mediator complexes have been studied for some time in yeast and more recently in mammalian systems. The mammalian complexes are homologous to those in yeast. Mediator complexes are huge. They contain from nine to twenty peptides and their mass ranges from 600 kDa to 1,000 kDa.

Surprisingly, different activators can interact with the same mediators. In addition, many of the subunits of different mediators are the same. However, aside from the shared subunits, the other subunits of each mediator are different. The mixing and matching allows for the needed variation in mediators. The size of the mediator complexes may reflect the need for interactions with different activators.

Apparently some of the mediators only act on chromatin and not naked DNA while others increase the rate of transcription on naked DNA. Repression complexes have been found to have a major role in the silencing of genes in eukaryotes. The complex of Ssn6 and Tup1 has been studied in some detail. These two proteins belong to a family of proteins represented in all eukaryotes studied. One Ssn6 molecule and four Tup1 molecules (all phosphoproteins) combine to form an elongated complex of 440 kDa. The repression induced by this complex is very efficient and may involve many genes. In Saccharomyces cerevisiae, the complex represses more than 150 genes.

The phenotype expressed by Tup1 and Ssn6 mutants is the result of inappropriate expression of these genes. The specific repression of a gene or set of genes can occur even if the activators are present. Generally, Ssn6-Tup1 binding to DNA is mediated by sequence specific DNA-binding proteins which bind to operator sites upstream from the gene. Although the mechanism of this inhibition is still not known two possibilities seem most likely: an alteration in chromatin structure and interference with the assembly of the transcription complexes.

The repression generally blocks genes that are essential to the survival of the cell under different conditions. Therefore, it is important to consider the mechanisms that remove the block. The most likely mechanism involves the inactivation of the DNA-binding proteins that attach to the Ssn6-Tup1 complex by blocking its transcription (if the protein has a high rate of turnover).

Another mechanism involves phosphorylation of the mediating proteins. The protein Crt1 becomes hyperphosphorylated when DNA is damaged and becomes unable to attach to the DNA. Similarly, when glucose levels are low, the protein Mig1 is phosphorylated and removed from the nucleus, thereby de-repressing the genes that are responsible for glucose metabolism.

Activation and Repression

A variety of covalent modifications of core histones have been found. Coactivatormediated acetylation and methylation of histones have been found to be part of the activation machinery of transcription. Other modifications include phosphorylation and ubiquitination. Many proteins that regulate transcription have histone acetyltransferase (HAT) or histone deacetylase (HDAC) activity or the ability to recruit the appropriate acetyl-active enzymes. HATs are involved in activation, whereas HDACs are involved in repression. Acetylation is thought to take place when a HAT is targeted to specific sites on the DNA by other protein molecules with which it forms a complex. This binding produces a selective disruption of the nucleosomal structure. In contrast, the HDACs are part of multicomponent complexes and may be required for full repression.

Many HATs are also coactivators of transcription and are recruited to promoters by activators after the latter attach to DNA. Conversely, certain histone deacetylases are transcriptional corepressor that are recruited to chromatin by repressor protein bound to DNA. Methylation of the histones leads to deacetylation during gene silencing. HDACs have been found to be associated with chromatin remodelling complexes with DNA dependent ATPase activity. Transcriptional repressors exert their effect by recruiting deacetylases to chromatin sites.

Ubiquitin has been found in the core histones H2A, H2B and H3.The ubiquinated histones are stable in vivo. In contrast to polyubiquitination, monubiquitination has been found to be linked to gene activation. The monoubiquination of H1 by TAF_{II} 250 was found to be a step necessary for specific gene expression during Drosophila development. TAFII, a component of the general transcription factor TFIID, was found to be a histone-specific ubiquitin-activating/conjugating enzyme.

Present information indicates that methylation, phosphorylation and acetylation of the core histones all play a role in the molecular events underlying activation and repression. For example, heterochromatin protein 1

(HP1) mediates gene silencing; loss of HP1 eliminates gene silencing and overexpression increases silencing. The silencing depends on a histone methylase. HP1 can bind with high affinity to histone H3 methylated at lysine. The results suggest that first a 'methyl marker' is placed on histone H3 and this marker is recognized by HP1.

Although HP1 is a component of condensed chromatin, its presence is not static. (Fluorescence recovery after photobleaching of a GFP-HP1 fusion protein shows that HP1 of murine T cells is highly mobile in euchromatin or heterochromatin and its activity is increased by activation of the T cells. The exchange correlates with the level of condensation of the chromatin and depends on the histone methyltransferase Suv39h. Maintenance of the heterochromatic (i.e., inert state) requires methyltransferases.

A specific antibody can detect methylation of lysine 9 in histone H3. The methylation was found to depend on a methyltransferase but also requires histone deacetylase. This appears to be followed by HP1 (or homologs) binding to produce silencing. Another mechanism appears to function in reverse and to involve phosphorylation. The phosphorylation of serine-10 facilitates acetylation of lysine-14. This suggests the existence of on and off states. In the off-state lysine-9 is methylated and lysine-14 is deacetylated and in the on-state serine-10 is phosphorylated and lysine-14 is acetylated. In addition, methylation of lysine-9 inhibits methylation of serine-10 and viceversa.

The interactions between acetylation and phosphorylation has been elucidated further by the isolation of a histone H3 serine-10 kinase complex. Its histone kinase subunit referred to as Snf1 was identified. Snf1 and the acetyltransferase Gcn5 were found to function in sequence. These observations provide further support to the idea that transcription is controlled by coordinated patterns of histone modification.

Transcriptional cofactors are also regulated by covalent modifications as indicated above. For example, the coactivator CREB-binding protein (CBP) when phosphorylated is recruited to certain promoters. In contrast, methylation of CBP blocks the binding to the cAMP response element binding protein (CREB).

Processing

Presently available information, primarily from in vitro experiments, allow the following reconstructions for the synthesis of mRNA in eukaryotes. The mRNA synthesized by the transcriptional machinery is much larger than the mature mRNA which is translated during protein synthesis. The large primary transcript has been called heterogeneous nuclear RNA (hnRNA). hnRNAs contain both coding and noncoding components. These are transcripts of the coding regions of the gene called exons and the noncoding regions called introns. The hnRNAs are associated with specific binding proteins to form heterogeneous ribonucleoproteins (hnRNAPs).

Twenty different hnRNAPs have been recognized so far, ranging in molecular weight from 31 to 120 kDa. After processing, the newly formed mRNAs are translocated to the cytoplasm. The resulting cytoplasmic mRNAs can then be translated. During processing the hnRNA is first spliced. In this process the non-coding sections of the hnRNAs are cut away and the portions of the spliced coding mRNA are joined together.

In eukaryotic cells, the nucleolus corresponds to the multiple ribosomal RNA genes in a specialized chromatin region which carries out ribosomal gene transcription, rRNA processing and ribosomal subunit assembly. In addition, the nucleolus is the site of processing and assembly of non-ribosomal RNAs. These include the maturation of the signal recognition particle RNA (SRP), tRNA, U6 small nuclear RNA, telomerase RNAs and some mRNAs.

Lampbrush chromosomes are meiotically paired chromosomes of amphibian growing eggs. They are very active in producing RNA and they form chromosome loops covered with large amounts of newly formed hnRNA bound to hnRNP proteins. Polytene (multistranded) chromosomes are present in salivary glands of fly larvae. The chromosomes grow to enormous size by replicating DNA, as many as a thousand times without cell divisions. The chromatids of these chromosomes are perfectly aligned so that different regions of the chromosomes can be readily distinguished morphologically. The dark bands seen in these chromosomes correspond to sites of condensed DNA.

Transcribed regions are non-condensed and are part of swollen puffs. These are the sites of transcription, demonstrable by growing the larvae in the presence of [3H] uridine, which is incorporated into RNA. The large puffs in Chironomus are called Balbiani rings. In addition to studies carried out with the light microscope, chromatin and nascent transcripts can be examined with the transmission electron microscope after spreading on a grid. The EM reveals the chromatin threads, RNA threads attached to it and granules. Together with other approaches, these EM studies examined the role of the various components in transcription. EM studies of Drosophila embryo, reveal nascent hnRNA transcripts, seen at active gene sites as smooth fibrils about 5 nm thick. These have been referred to as perichromatin fibers.

Splicing

Most eukaryotic genes the primary transcript is much larger than the mRNA. This is because the DNA has coding regions (exons) and in-between regions (introns). The segments corresponding to the introns are removed by a process known as splicing. Splicing must involve various distinct tasks. Every exon must be located in the primary trancript, then the introns must be removed and the appropriate coding sequence must be joined precisely and in the correct order. We have seen that splicing requires hnRNPs, snRNPs and splicing factors.

Normally, splicing is cotranscriptional, however, it can occur postranscriptionally. In one of the early studies this was demonstrated when pre-mRNA from the b-globin sequence was synthesized in vitro using bacteriophage RNP. After injecting the pre-mRNA into Xenopus oocytes, the mRNA was recovered and was found to have been accurately spliced. Clearly, transcription and splicing can take place as two separate processes.

The demonstration that transcription and splicing can occur independently suggests an experimental approach to determine what components are needed for processing. Synthetic pre-mRNA could be added to extracts and the processing of the RNA followed by gel electrophoresis. By adding them one at a time to a partial mixture, the role of each protein fraction could be tested for its ability to produce either mRNA or partially spliced products. A number of necessary factors have been identified in this way. This approach revealed the need for interactions of hnRNA with the spliceosomes and each of their components.

Many aspects of the biochemistry of splicing have been examined. Splicing of some prerRNAs, mitochondrial and chloroplast pre-mRNA, and tRNA do not require a spliceosome. These RNAs self-splice by a similar mechanism in which the RNA itself functions as an enzyme. Segments of the same mRNA can be spliced in alternative ways, giving rise to different proteins from the same gene. This has been referred to as alternative splicing. One third of human genes are thought to give rise to alternatively spliced mRNA.

Splicing of hnRNA requires spliceosomes, large ribonucleoprotein assemblies containing a large number of proteins and small nuclear RNAs (snRNAs). Neubauer et al. characterized the entire mammalian spliceosome complex using mass spectroscopy in conjunction with DNA and protein sequences data bases. All components of the spliceosomes, including small nuclear ribonucleoprotein particles (snRNPs) and splicing factors, are found in the nucleoplasm.

The snRNPs of the spliceosome complex include U1, U2, U5 and U4/U6. During the assembly of the spliceosomes, the pre-mRNA and the RNA components of the snRNPs interact. All sliceosomal proteins share common structure and mechanisms of assembly. Mammalian U1 snRNP consists of ten different proteins and an RNA consisting of 165 nucleotides. Seven of the proteins, the Sm proteins are shared by all U snRNPs. The other three are specific to U1. The snRNPs of the spliceosome bind sequentially to the pre-mRNA. U1 is bound first followed by U2, U2/U6 and U5.

A reconstruction at 10 Å resolution using cryo-electron microscopy allows assigning seven Sm proteins as doughnut shaped elements and arriving at a model for U1 snRNP which agrees with a circular structural model of Sm protein arrangement. The localization of hnRNAPs and snRNAPs in the cell can be studied using immunofluorescence. In the case of splicing factors, there are 20 to 50 spots of high fluorescence. Similar results have been obtained when the site of snRNA was identified using hybridization techniques with the appropriate DNA probe.

With this technique, two complementary single strands of DNA or RNA can be annealed to form a helical structure. This can be carried out with two complementary DNA strands or a DNA and an RNA strand. One of the two can be used as a probe, for example, if labelled with a radioactive isotope, recognized by autoradiography). The regions in which the splicing factors are present in high concentration have been called speckles and they are about 0.5 to 1.5 mm in diameter. In nuclei of some cells, the snRNP proteins are localized in small structures first observed in 1903 by Ramon-y-Cajal, called the coiled bodies because of their appearance as coiled webs.

The EM reveals that the speckles contain interchromatin granule clusters (IGC) in which each granule is approximately 20 to 25 nm in diameter. Perichromatic fibers extend from the IGCs and are also found in the nucleoplasm. We have seen that these fibers represent nascent transcripts. Although transcription occurs at the sites of perichromatic fibers, splicing factors can be found predominantly in the IGCs and not in the perichromatic fibers. The IGC could therefore represent a storage site for splicing factors from which they would have to be recruited.

The translocation of splicing factors to the sites of transcription has been shown in the case of viral and plasmid transcription. The translocation of splicing factors to the active sites and their exit when the sites become inactive, has also been demonstrated. A short amino acid motif of some splicing factors (SR-proteins), is sufficient to target the proteins to the speckles. This motif is present in a serine-arginine rich domain, the RS-domain that is heavily phosphorylated.

Four kinases have been shown to act specifically on SR-proteins and at least two of them have been shown to be responsible for the intranuclear localization of the proteins. Furthermore, in at least some cases, SR-proteins and some snRNP proteins have to be phosphorylated to function in spliceosomes. In addition, dephosphorylation is probably required for the splicing reaction and the disassembly of spliceosomes.

An involvement of phosphorylation in recruitment to the transcriptional sites would imply that splicing factors must be released from the IGCs and then returned when no longer needed. The return to IGCs depends on phosphatase activity. Similarly, overexpression of the a kinase that localizes to speckles, becomes more diffusely distributed when phosphorylated and overexpression of the kinase displaces SR-proteins from the ICGs.

Macromolecular Traffic

Proteins are synthesized in the cytoplasm and mRNAs are transcribed and processed in the nucleus. Therefore, the nuclear envelope exerts an all important role in transcriptional control (by selecting the proteins entering the nucleus which may have a role in gene expression) and post-transcriptional control (by selecting the mRNA molecules leaving the nucleus). Proteins need nuclear localization signals (NLSs) to be translocated into the nucleus. RNA export also requires some sort of signal sequence. However, many aspects of the role of the nuclear envelope are still to be explored in detail and, except for a few cases, this aspect cannot be clearly evaluated with the information presently available.

The general role of the nuclear envelope in regulating macromolecular traffic is an important component of gene expression. This can be demonstrated with an example: the role of translocation through the nuclear envelope in the regulation by steroid hormones. The hormone binds to the glucocorticoid receptor (GR) protein present in the cytoplasm. The binding induces a dissociation of the receptor subunits. The receptor subunit binding the hormone and now released from the complex can enter the nucleus and bind the hormone response element (HRE) in the DNA, which can now activate transcription of the downstream DNA sector.

Interactions

As we saw, the production of mRNA and its translocation

from the nucleus to the cytoplasm occurs in a number of steps which include transcription (initiation, elongation and termination), capping of the 5' end, removal of introns, cleavage at the 3' end, polyadenylation and after release from the chromatin, the translocation into the cytoplasm. These processes are in series and tightly coupled physiologically and physically. In part these connections are arrived at by tethering to each other the machines responsible for each step.

For example, RNA polymerase II and the RNA processing machinery are coupled. Similarly, the coupling of transcription to splicing is produced by tethering splicing factors to the transcriptional machinery. Likewise, the splicing of mRNA and its translocation are also coupled by attaching the nascent mRNA to a export factor Yra1p and transferring it to the receptor (TAP/Mex67) and its delivery to the nuclear pore complex.

Noncoding RNA

A variety of RNAs do not function as mRNA, tRNAs or rRNAs. These have been referred to as noncoding RNAs (ncRNAs) (small RNAs in bacteria). NcRNAs have been found to have a role in transcriptional regulation, chromosome replication, RNA processing and modification, mRNA stability and translation, as well as protein degradation and translocation. The sizes of ncRNAs varies depending on their function. Those associated with development in the nematode Caenorrhabditis elegans, Drosophila and mammals have been found to be 21 to 25 nucleotides in length. The translational regulators in bacteria are 100 to 200 nucleotides in length and those involved in gene silencing in eukaryotes are larger than 10,000 nucleotides.

RNA Localization

RNA localization at specific sites is involved in

differentiation and embryogenesis. In addition, it determines cellular asymmetry and cellular polarity, and has a role in synaptic plasticity. It produces a high concentration of protein at a specific cellular locations (e.g., actin at the leading edge of cells) and segregates RNAs to specific sites (e.g., in regions of a neuron). During embryogenesis, it has a role in the production of morphogens and the formation of cell lineages by restricting the presence of RNA to certain cells.

The localization results from mechanisms which include vectorial transport out of the nucleus, cytoplasmic transport to targeted sites by the cytoskeleton, local entrapment at particular cytoplasmic sites and local stability accompanied by the general degradation of the RNA at other sites.

Targeting sequences in the mRNAs 3'-untranslated region (UTR) serve as signal sequences for their localization. In other cases, RNA-binding proteins have a role in the targeting. The binding of these proteins may associate the RNA with a variety of other proteins to produce large ribonucleoprotein (RNP) particles which are transported to their targets. When the RNA arrives at its destination it is anchored by proteins or in some cases other RNA molecules.

The regulation of the translation of localized mRNAs has a central role in synaptic plasticity and hence in memory and learning. Synaptic activation through the Nmethyl-D-aspartate receptor (NMDAR) can activate the translation of mRNAs. For example, the synthesis of the b subunit of the calcium-calmodulin-dependent kinase II, required for learning and memory, is stimulated by the brain-derived neurotrophic factor (BDNF) which functions in synaptic plasticity.

Many of the localization proteins also have a nuclear function suggests that these proteins have multiple function. In addition, it has been shown that certain localization proteins may associate with specific transcripts.

TRANSLATION

In many cases, regulation of gene expression occurs during transcriptional processes. However, many experiments indicate that regulation of gene expression can also occur after transcription, that is, posttranscriptionally. Some of this regulation occurs during the process of translation. Three phases are involved in translation: initiation, elongation and termination.

Initiation involves the formation of a complex between the initiation factor eIF2, GTP and tRNA-Met (1), with 40S subunit of ribosomes complexed to eIF-6 and eiF-3 [indicated as 3 and 4C respectively] (2). This complex (3) can then bind to the m7G cap of mRNA at the 5' end of the mRNA (4), where it scans the mRNA in the 5'-3' direction until it reaches the AUG initiation codon (5). The factors are then released (6), GTP is hydrolyzed and the 60S subunit joins the mRNA where the Met-tRNA is held at the initiation codon, to form the initiation complex (7).

Different initiation factors are part of complexes which act at different steps. At least 24 of the eIF polypeptides have been cloned. Those of the eIF-1 class are thought to be involved in the pleiotropic stimulation of the assembly of the translational complex. Elongation requires a triplet code where a combination of three bases (a codon) generally corresponds to one amino acid.

Each tRNA species corresponds to a single amino acid and has a triplet complementary to that present in the coding mRNA (an anticodon). In this model (starting with 1), E is the exit site, P is the peptidyl-tRNA site and A is the site where the incoming aminoacyl-tRNA will bind. EF, followed by the appropriate number or symbol, indicates the involvement of an elongation factor. The amino acyl-t-RNA attaches to the codon in site A (2 and 3).

The tRNA-peptide at site P is transferred to the aminoacyltRNA in site A and a peptide bond is formed between the aminoacyl-tRNA and the peptidyl-tRNA. In step 4, the 3 end of the deacetylated tRNA is now shifted to the E site without moving the anticodon. The peptidyl tRNA chains grows by one amino acid without moving. Translocation (step 4 step 1) moves the mRNA by one codon so that a new codon is at the A site. For the step immediately following initiation, the same model is proposed. However, in this case the Met-tRNA acts as the peptidyl-tRNA shown in this model.

Termination involves one of the stop codons (UAA, AGA and UAG) and a release factor, RF. RF binds GTP, recognizes the stop codon and induces the hydrolysis of the aminoacyl linkage and the GTP. For simplicity, the previous discussion ignored in most respects the ribosome. However, the ribosome represents the translational engine. In fact, protein synthesis may be carried out by ribosomes in the absence of factors or GTP under non-physiological conditions, indicating that the necessary machinery is totally within ribosomes.

In bacteria, the ribosome is about 2.5 MDa and is constituted by more than 50 different proteins and 3 RNA molecules representing over 4500 nucleotides. The two subunits are denoted by their sedimentation constant, S. Peptide synthesis involves complex interactions in the ribosome. The smaller subunit (in bacteria 30S, 40S in eukaryotes) of the ribosome binds mRNA and the anticodon end of the tRNAs, the larger subunit (in bacteria 50S, 60S in eukaryotes) binds to the amino acid end of the tRNAs and catalyzes the formation of the peptide bond. After the peptide bond is formed, elongation factor G (EF-G) binds to the ribosome. The peptidyl-tRNA is then moved by one codon of the mRNA.

Clearly, the process requires movement. Escherichia coli 70S ribosomes in the various stages were examined by cryo-electron microscopy and three-dimensional reconstructions. The binding of EF-G and the GTP hydrolysis produce a ratchet-like rotations of the small subunit relative to the large subunit. The translocation occurs in two steps. First GTP binds to EF-G, followed by rotation of the subunits and opening of a channel occupied by mRNA. This is then followed by GTP hydrolysis and advance of the mRNA/(tRNA) complex in the direction of the rotation of the 30S subunit. In addition, many other conformational changes take place within the two subunits.

Coordinate Expression

The transcription of many genes is frequently coordinated. DNA-microchip technology has allowed examining in detail the coordinate expression of related genes in eukaryotes over several hours. In this so-called synexpression, there is a close correlation between function and expression. However, in contrast to the operons of bacteria that are transcribed as polycistronic mRNAs, synexpression takes place by the coordinate activation of genes in different chromosomes.

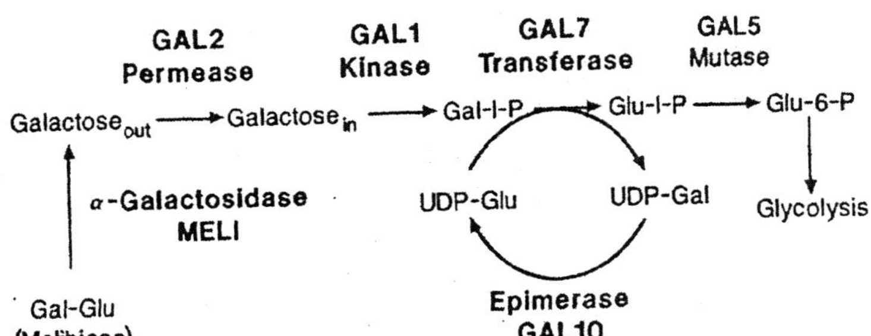

Figure 1. Pathway of galactose utilization

The sequence of transcriptional events accompanying the metabolic shift from fermentation to respiration, was studied using DNA microarrays containing almost all of the genes of Saccharomyces cerevisiae. The shift produced the coordinate enhancement and decline in the transcription of hundreds of genes. The control of galactose metabolism in Saccharomyces cerevisiae by galactose represents a small sample of how some of these coordinated systems operate. In yeast, galactose is generally metabolized after conversion to glucose-6-phosphate in a pathway depicted in Fig. 1.

In this representation, the genes coding for specific enzymes are listed on top of the enzyme. In the schematic representation of the pathway below these listings, *Gal* and *Glu* correspond to galactose and glucose respectively, and P is phosphate. Melibiose is first hydrolyzed to its two component monosaccharides extracellularly. After transfer into the cells mediated by a permease, Galactose is then metabolized in four separate steps (2-5) to produce Glucose-6-phosphate which is subsequently broken down through the reactions of the glycolytic pathway, as indicated. Except for the *Gal5* which is unregulated, the expression of the genes of this pathway is highly regulated. The presence of galactose induces the complete pathway.

The model of gene regulation which explains most of the data presently available, is represented in Fig. 2. In this figure, each line represents a different chromosome. The various GALs correspond to the genes indicated in Fig. 1. *Gal1, Gal7* and *Gal10* are present in the same chromosome and have separate promoters so that their transcription is separate. *Gal2* and *Mel1* are present in other chromosomes.

The expression of all these genes is activated by a protein encoded by *Gal4,* as indicated by the stippled arrows. The *GAL4* protein binds to the DNA upstream of these genes. Regulation, however, depends on an additional gene. *Gal80* encodes a protein which binds directly to *GAL4* protein, inactivating its effect, as indicated by the stippled bar. Inducer binding of *GAL80* protein prevents its inhibition of *GAL4*. Empty arrows in this figure indicate the direction of the transcription, solid arrows indicate the production of the protein indicated. The DNA sectors which bind the *GAL4* protein, are the so called upstream activation sequences which correspond to enhancers of higher eukaryotes.

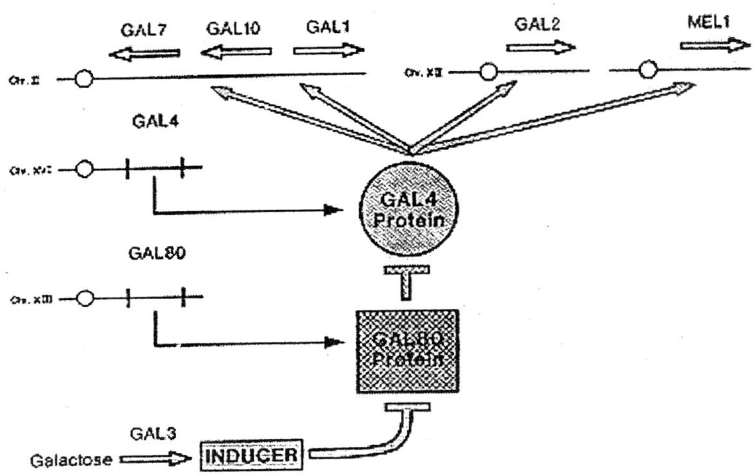

Figure 2: Components of the GAL gene regulatory circuit

The gel electrophoresis patter of the UASG of the *Gal*+ DNA is compared to that of a Gal- mutant. The sectors protected by binding to the GAL4 protein are missing from the gel. The various DNA sectors were identified using radioactive probes complementary to the DNA sequences adjacent to UASG.

The GAL4 protein has been found to contain various domains. One of these binds GAL80 protein. Another contains a nuclear localization sequence (NLS) which allows it to enter the nucleus. A Zn-finger domain binds DNA and a very acidic domain acts in transcription activation. The GAL4 protein is present constitutively.

GAL80 binds GAL4 only in the absence of an inducer. The interaction has been shown with UASG-DNA bound to Sepharose beads. In the presence, but not in the absence of GAL4 protein, GAL80 attaches to the beads.

Regulatory Devices

Factors which regulate transcription, recognize elements called response elements (RE) which can be present in promoters or enhancers. For example, the glucocorticoid hormone receptor acts by binding to the glucocorticoid response element (GRE). Activation of a gene may involve the recognition of a promoter or enhancer by a transcription factor, so that transcription can be initiated. In some cases, there are several genes with common REs. Conversely, a single gene may be regulated by several REs, each corresponding to a different promoter.

In contrast, repression may result from binding the target DNA sequence by a repressor protein, blocking the accessibility of the promoter to the transcriptional machinery. The basic transcriptional machinery frequently requires sequence specific factors which bind to the protomoter/enhancer elements. A family of proteins have been identified which act as coactivators of transcription. They do not bind with the promoter elements but become part of the transcriptional initiation complexes. They connect the specific DNA-binding factors to the basic transcriptional machinery. These include the CREB (CREB=cAMP regulated enhancer binding) binding protein. These proteins are the target of many regulative effects, such as the regulation by signal induced protein kinases. However, transcription factors and other proteins are not alone in regulating eukaryotic transcription.

Histones and their organization in nucleosomes are of fundamental importance. For transcription to occur, the remodeling of repressed chromatin must take place. Chromatin transitions are apparently carried out by so-called chromatin remodeling machines (CRMs). All these complexes contain subunits homologous to DNA-dependent ATPases. They all carry out ATP-dependent perturbation of chromatin. Their mode of action differs. The nucleosome remodeling factor (NURF) facilitates the ATP dependent binding of transcription factors to the chromatin.

The chromatin accessibility complex (CHRAC), a complex of five proteins, contributes to nucleosome spacing and makes chromatin more accessible. The ATP-dependent chromatin assembly and remodeling factor (ACF) consisting of two subunits (Acf1 and ISWI) contributes to the assembly of nucleosome arrays and facilitates the binding of activators to chromatin. The ISWI subunit is related to helicases and is responsible for the ATPase activity in which 2 to 4 molecules of ATP are hydrolyzed per base pair during nucleosome formation. The nucleosome remodeling involves the histone tails of the nucleosomes. ACF was found to assemble nucleosomes in organized arrays.

SWI-SNF and related complexes act in restructuring chromatin and are thought to play a role in transcriptional activation and the control of the cell cycle. In a model system, SWI-SNF has been found to displace nucleosomes along the DNA to another position. The functioning of the CRMs is still being evaluated; each is thought to act in a distinct manner.

RNA polymerase II has been been found to be associated with the gene regulator proteins SWI/SNF that are active in chromatin remodeling as discussed above. The CREB binding protein CBP is also involved with the action of the polymerase. Apparently its role is that of an adaptor for many transcription factors and its action is thought to be mediated by its recruitment of histone acetyltransferases (HATs) to the promoter. In addition, CBP itself has been found to have HAT activity. The acetylation of histones has a role in gene expression.

Phosphorylation of histones plays a role in gene activation. Growth factor activation via mitogen activated proteins (MAPs) phosphorylates H3. The phosphorylation corresponds to the H3 histone of fos and myc genes of stimulated cells, as detected by an antibody specific for phosphorylated Ser-10 of H3. myc, fos and jun genes are protoncogenes with a role in the regulation of cell growth. When overexpressed or hyperactivated they cause uncontrolled cell proliferation. Stimulation of the Ras-MAPK signaling pathway in oncogene-transformed cells also results in increased amounts of phosphorylated histone H1.

Ras-GTP binding proteins are responsible for transducing the signal from receptor tyrosine kinases to the nucleus to stimulate proliferation. The treatment of mouse fibroblast with various growth factors or protein synthesis inhibitors results in the transient phosphorylation of H3 and HMG14. Apparently, the growth factors trigger a kinase cascade (the ERK pathway) which activates p42/p44, an activity mediated by GTPbinding proteins Ras and Raf.

In mammalian cells, the phosphorylation of H3 associated with transcription is unique to a small set of genes. How phosphorylation of the histones can affect gene expression is still an open question. Phosphorylation may act by adding to the histones negative groups that disrupt the interaction of the H3 tails and the negatively charged DNA backbone. Alternatively, phosphorylation could promote gene activation by serving as a site for recruitment of transcription factors. Some transcription factors contain SH2 domains which correspond to phosphotyrosine binding domains.

Post-transcriptional Regulation

Posttranscriptional regulation includes a variety of controls including the degradation of mRNA and proteins. In some cases, many reactions, usually in the same pathway, are regulated by the same mechanism. In other cases, only one or a few related genes are regulated by the same mechanism.

Regulation of Translation

The global control of protein synthesis is necessarily linked to physiological conditions and in particular to the availability of the supply of the amino acids. The control of the synthesis of specific proteins also takes place. The rate limiting process is frequently the initiation of translation, so that in these cases regulation occurs at this step. The most frequent mechanism for the regulation of protein synthesis is the phosphorylation and dephosphorylation of translational components, mostly initiation (eukaryotic initiation factor, eIF) or elongation factors (eukaryotic elongation factors, eEF). In some cases the phosphorylation activates the factor, while in other cases phosphorylation can blocks initiation.

Genetic manipulation of one of the eiFs (eIF-4E) has shown that this initiation factor is involved in the regulation of growth and cell proliferation. eIF-4E is the protein responsible for the recognition of the m7GpppN cap at the 5' end of almost all eukaryotic mRNA and it facilitates the melting of the secondary structure of 5'UTRs. Cells transfected with vectors containing the gene for this protein grow rapidly and form tumors. Conversely, the expression of RNA antisense for the mRNA of eIF-4E (i.e., the RNA complementary to the normal mRNA, which would therefore inactivate the latter by specifically binding it) decreases protein synthesis and cell proliferation. The phosphorylation of this factor enhances its activity and promotes protein synthesis.

A family of regulators that bind eIF4E were recently discovered. PHAS/4E-BP1 inhibits the eIF-4E-dependent mRNA translation by blocking the binding of eIF4-E to eIF4-G. The latter interaction is required for efficient interactions of the 40S ribosomal subunit to the mRNA. The effect of PHAS/4E-BP1 is most pronounced in the translation of mRNAs with highly structured 5'UTRs rather than being global.

The regulation by several agents (e.g., insulin-like growth factor and interleukin 2) promotes the dissociation of PHAS/4EBP1, whereas those that increase cAMP increase the binding. The effect of insulin is most likely to be in response to a phosphorylation of PHAS/4E-BP1 and the subsequent release of eIF4E. The phosphorylation and dephosphorylation of PHAS/4E-BP1 is the most likely mechanism for the action of other agents as well. The effect of eiF-4E and the control exerted by some genes is very general. However, the regulation may be much more specific, sometimes due to the presence of highly structured 5'UTRs in the mRNA. However, the control of individual genes and that of an entire pathway may be very similar because a single gene may affect a variety of other genes.

Regulation of Specific Genes

The presence of specific mRNA and its availability to the translational machinery has a role in gene expression. In some cases, a pattern of repression-derepression similar to that occurring at the gene level has been found. In these cases, repressors or activators bind to specific sectors of the mRNA, blocking or facilitating translation. In other cases, the regulation is more complex.

Regulation involves the phosphorylation of the a-subunit of eIF2. The phosphorylation downgrades translational initiation either in general or at specific steps. The eIF-2 kinases have their kinase catalytic domain in common. However, they contain unique regulatory regions adjacent to the catalytic domain, permitting regulation by different physiological signals. The phosphorylation of eIF-2a was first found in reticulocyte lysates deprived of hemin. The absence of hemin activated the eIF-2a kinase, the heme-regulated inhibitor (HRI).

The double-stranded RNA-activated kinase (PKR) is constitutive in reticulocytes and inducible by interferon in other cells. PKR is activated by double stranded (i.e., viral) RNA in the presence of ATP. This kinase is part of the cell's defense against viral infection. Association of PKR with double stranded RNA results in autophosphorylation and consequent activation. In yeast (Saccharomyces cerevisiae), the general control nonrepressible kinase (GCN2) is activated by uncharged tRNA during amino acid starvation and purine shortage. There are also indications that other eIF-2 a kinases exist and may play a regulatory role.

Gene specific translational control by the phosphorylation of eIF-2 a is still not well understood. The regulation of GCN4 by GCN2-protein kinase depends on the relative rates of the translation of upstream open reading frames (uORFs) in the GCN4-mRNA.

Regulation of iron : Iron is a vital component of living cells. Hemoglobin contains iron and so do the cytochromes which are responsible for the electron transport in mitochondrial oxidative phosphorylation. Not surprisingly, various components involved in Fe storage or transport, or the synthesis of Fe-containing compounds, are regulated by the availability of Fe. Ferritin is an intracellular protein of animal, plant, fungal and bacterial cells responsible for binding and storing Fe. Aminolevunilic acid (ALA) is one of the precursors of porphyrins needed for cytochromes and hemoglobin. Transferrin is a protein which carries iron in the blood and is taken up by cells through endocytosis mediated by the transferrin receptor (TfR).

Accordingly, the presence of Fe in the cell favors the synthesis of ferritin and ALA synthase, but inhibits that of the TfR responsible for the uptake of Fe and which is needed in lesser amounts at high Fe concentrations. The mRNAs corresponding to the proteins which are needed in larger amounts in the presence of iron (ferritin, ALA), have a 35-nucleotide region (iron response element, IRE) in the 5' untranslated region of the mRNA.

A ferritin repressor (FRP) has been shown to bind to the IRE in vitro. In contrast to the case of ferritin, the TfR is needed in larger amounts in the absence of iron to make up for its low concentration. The TfR-mRNA contains several copies of the IRE, similar but not identical to that of ferritin, in the 3' untranslated region. In this case binding of the repressor stabilizes the mRNA, so that the synthesis of the transferrin receptor is enhanced.

Tubulin regulation : The regulation of the synthesis of the tubulin, the subunits of microtubules, follows a mechanism in which the stability of mRNA depends on events accompanying translation. Microtubules are structures which participate in a spectrum of cellular functions. Together with intermediate filaments and actin, they are basic structural elements determining cell shape and growth.

Furthermore, they participate in vital cell functions, such as cell movement, the formation of the mitotic spindle and the movement of chromosomes during mitosis. The basic units of tubulin are the heterodimers of a and b tubulin. Microtubules are in rapid, dynamic steady state with the free tubulin dimers. During cell differentiation, one or more of the 6 to 7 genes that encode either subunit are activated.

However, the rate of synthesis of tubulin needed to maintain the appropriate concentration is determined by the concentration of free tubulin dimers by a negative feedback mechanism at the posttranscriptional level. The control of the tubulin synthesis by the tubulin level can be demonstrated by microinjection of tubulin into cultured cells, or in experiments in which the cells are treated with drugs, such as colchicine (which depolymerizes microtubules and therefore increases the concentration of free tubulin). Increased free tubulin dimers decrease the amount of tubulin being synthesized.

The regulation is posttranscriptional: the presence of a nucleus is not required. Analyses of tubulin mRNA levels show that the regulation can be accounted for by decreases in mRNA stability. The stability of the tubulin-mRNAs is regulated by the level of the free heterodimer which apparently depends on the presence of four amino acids in b-tubulin.

As few as 13 nucleotides artificially added at the 5' end of mRNAs coded for another protein, were found to confer b-tubulin sensitivity. In addition, the regulation occurs only when the translation goes past the first 41 codons of the mRNA, as indicated by experiments carried out with truncated mRNAs, and requires the association of the mRNA to ribosomes to form polysomes. These results suggest that the four terminal amino acids of b-tubulin bind to the free tubulin heterodimer as it emerges from the large ribosomal subunit during translation. This binding would activate a nuclease.

Synthesis of Amino Acids

GCN4 is a gene which regulates the biosynthetic pathway of amino acids in yeast. It activates more than 30 genes coding for enzymes involved in the biosynthesis of 10 different amino acids. *GCN4* expression increases during amino acid starvation. Conversely, when amino acid levels increase, *GCN4* expression decreases. During these changes, the level of *GCN4* mRNA remains the same. The control is therefore translational.

The control by *GCN4* involves a subtle interplay of translational events, in which the initiation complex and the *CGN4* mRNA untranslated 5 portion of 590 bases play a role. The initiation complex contains Met-tRNA, where Met is the amino acid which initiates peptide synthesis. The mRNA region contains four small open reading frames (ORFs), referred to as ORF1 to ORF4. An open reading frame is a sequence of codons with a start and a terminator codon which could potentially code for a polypeptide.

As shown in Fig. 3, the 40S subunit of ribosomes scans from the 5' end of the mRNA. They initiate translation when they arrive at either ORF-1 or ORF-2. The complete translational machinery, including an assembled ribosome, is bound at this site. At termination of the reading of each ORF, the translational machinery disassembles, but the 40S ribosomal subunit continues scanning in the 5' to 3' direction. When amino acids are present in abundance, the eIF-2.GTP.Met-tRNA complex reforms readily and reinitiation starts at ORF4.

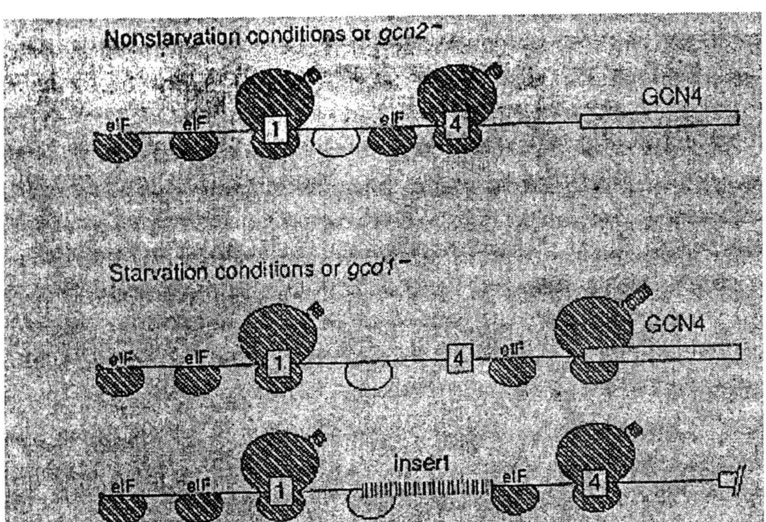

Figure 3: Model for *GCN4* translational control

However, the scanning of the 40S subunit cannot recognize the AUG initiation codon of GCN4, because the complex does not have enough time to reassemble after leaving ORF4. Therefore, GCN4 cannot be translated. Without GCN4, the synthesis of new amino acids is inhibited. When eIF-2.GTP.Met-tRNA is formed slowly (low level of amino acids, line 2), the scanning ribosome bypasses ORF-3 and ORF-4 because the initiating machinery fails to assemble in time. However, the complex has enough time to reform and initiate translation at the AUG codon of GCN4. The increased, newly synthesized GCN4, will therefore activate the synthesis of new amino acids.

This mechanism, based on the rate of separate translational events, is thought to allow eightfold fluctuations of GCN4 between starvation (high GCN4) and high amino acid concentration (low GCN4). Line 3 of Fig. 3 illustrates one of the key pieces of evidence for this model. In accordance with the model, when an extra piece of RNA is inserted before ORF4, CGN4 is not produced, even under starving conditions. This is because the initiation complex has enough time to reform at ORF4. GCN4 is also controlled by the general control nonrepressible kinase (GCN2) which phosphorylates eIF-2a.

The positive regulatory role of GCN2 on the amino acid synthetic pathway is one of the few cases of translational control of the expression of a gene now partially understood. GCN2 is activated by uncharged tRNA during amino acid starvation. The phosphorylation of eIF-2 stimulates the translation of GCN4-mRNA. In the case of GCN4-mRNA in yeast, the regulation is through the uORFs. The phosphorylation of eIF-2 reduces the activity of eIF-2. The resulting delay of the assembly of the translational-ribosome complex at the uORFs downstream from uORF1, permits the translation of the GCN-4 protein.

RNA EDITING

After transcription, certain RNAs are edited, i.e., nucleotides are deleted, inserted or replaced. The most far-reaching alterations are the consequence of editing mitochondrial RNA and, in one extreme case, as much as half of the mRNA is changed. Generally, the editing in nuclear transcripts is much more limited and may involve a single base. In mRNAs coded by nuclear genes, editing precedes splicing.

Frequently editing is necessary for the RNA to function; the editing is required to produce initiation and termination codons or eliminate internal frameshifts. Frameshift mutation removes a single base in the mRNA and would lead to misreading of the rest of the mRNA (the codons are base triplets, removing one base shifts the reading so that subsequent codons would be different).

Some of the editing is needed to maintain a necessary, three-dimensional structure. For example, in tRNAs, function depends on a precisely formed three dimensional cloverleaf conformation. In other cases, the editing is to change an internal amino acid in the corresponding protein. The occurrence of these events has been detected in experiments in which the genomic sequence did not correspond to the cDNA sequence (mirroring that of the RNA rather than the gene).

Recently, editing in two mammalian mRNAs has been shown to permit the production of two distinct proteins from the same gene. This type of editing may be under developmental or physiological control. One case corresponds to a plasma lipoprotein, apolipoprotein B (apoB), which occurs in two forms: apoB-100 (of 512 kDa) and apoB-48 (of 241 kDa). The two apoB forms are involved in the transport of cholesterol and triglycerides in the blood. Another case corresponds to the glutamate-gated channels present at synapses.

ApoB-100 is produced mostly by the translation of the mRNA encoded by the genomic sequence in the liver. ApoB-48 is produced by the translation of an mRNA from the same gene, but edited to produce the shorter protein by substituting a stop codon (UAA) for a glutamine codon (CAA). The editing is tissue specific and is under physiological control, responding to nutrients and hormones.

The glutamate receptors (gluR) fall into three separate subgroups, depending upon the activating pharmacological agonists: the N-methyl-D-aspartate (NMDA), the a-amino-3-hydroxy-5-methylisoxazole-4-propionic acid (AMPA) and kainate (KA) classes. So far, 28 gluR genes have been identified and others are known to exist. The AMPA receptors differ in the ion selectivity. The mRNA of the AMPA and KA classes are edited by substitution of a glutamine codon (CAG) with an arginine codon (CGG) in a portion coding for the channel segment of the protein. Apparently, the presence of arginine blocks the passage of Ca^{2+}. The control of Ca^{2+} permeability, in turn, is likely to have important physiological consequences (for example, Ca^{2+} is a second messenger).

The editing is tissue specific within the nervous system, and it has been proposed that it plays a role in the regulation of expression during the establishment of excitatory circuits in the developing nervous system. Several studies have suggested that glutamate receptors are involved in learning and memory. In at least some cases, the enzyme involved in the editing recognizes a double strand locus in the RNA formed by the hybridization of sequences in the intron to complementary sites in a nearby exon.

RNA Interference

The events known as RNA interference have gained prominence in our understanding of posttranscriptional regulation. The best understood phenomenon is the RNA interference involving double stranded RNA (dsRNA). dsRNA degrades homologous mRNA with the same sequence as one of the strands of dsRNA. This process proceeds through the production of short interfering RNAs (siRNAs) and is thought to provide a mechanism to remove defective RNA and protect cells from virus that use dsRNA to replicate, transposons and repetitive sequences.

Gene silencing via RNA interference also has been shown to have a role during the development of the nematode Caenorhabditis elegans, Drosophila melanogaster and other animals through the production of small temporal RNAs (stRNAs; also known as microRNAs, miRNAs), which contain sequences complementary to specific target mRNAs. Unlike dsRNA, stRNAs are encoded by the genome of the developing cells and repress but do not degrade the target mRNA.

Despite their different origin and probable role, there are many similarities between the siRNA and the stRNA systems. Both act through small RNAs produced from longer RNAs. Both involve the ribonuclease Dicer, which produces the smaller pieces of RNA. However, whereas in siRNA Dicer acts on dsRNA, in stRNA, it acts on RNA. coded by the genes let-7 or lin-4).

Following their formation, the siRNAs are incorporated in the RNA-induced silencing complex (RISC) which targets homologous RNA for degradation. The process proceeds in steps: (1) an ATP dependent processing in of the double-stranded RNA into siRNAs, followed by (2) incorporation of siRNAs into a 360 kDa protein-RNA complex and (3) an ATPdependent unwinding of the siRNA duplex, and finally (4) an ATP-independent recognition and cleavage of the RNA target. A similar system has been found in plants. stRNAs block the translation of mRNAs by pairing with the 3' untranslated regions (UTR) of these RNAs. They have also been found in plants. They are thought to control gene expression by base pairing with mRNAs. MiRNAs are part of a 550 kDa ribonucleoprotein complex containing many molecules of miRNAs.

REFERENCES

Berg J, Tymoczko J.L., Stryer L., *Biochemistry*, 6th ed., San Francisco: W. H. Freeman, 2006.

Robert J., *Brooker Genetics: analysis and principles*, 2nd edition, New York: McGraw-Hill, 2005.

Mohamed Ouhammouch, Robert E. Dewhurst, Winfried Hausner, Michael Thomm, and E. Peter Geiduschek, "Activation of archaeal transcription by recruitment of the TATA-binding protein", *Proceedings of the National Academy of Sciences of the United States of America*, 2003.

6

Cell Motility

Cell motility describes the methods that single-celled organisms use to both move in their environments and move things internal to the cell. Different types of cells use a variety of strategies to seek food, produce energy, avoid death and fulfill roles. In multicellular organisms, cell motility enables single cells and groups of cells to migrate in the embryo to the appropriate locations in the body. Unfortunately, this same process critical to physical development becomes deadly when uncontrolled, leading to cancer and the metastasis of cancerous cells.

Most cells in complex animals are stationary. However, cell motility may also be demonstrated in morphology: muscle cell contraction, nerve axon elongation, cell-surface protrusions, even the dividing of cells during mitosis. Inside cells, much more dynamism in cell motility is observed: cytosol streams, vesicle transport systems, and the separation of chromosomes and creation of RNA. Everything in a normal cell is carefully controlled. ATP fuels all cell movements, along with the action of proteins that convert ATP's energy.

The cytoskeleton, a changeable system of fibers in the cytoplasm, supports the cell membrane and provides "tracks" for organelles and other elements to move through the cytosol. This system is constantly rearranging itself to suit new processes. This system is a secondary source of cell motility.

CELL MOVEMENTS

A wide variety of cell movements has been characterized by biologists and biophysicists to varying degrees of molecular and mechanistic precision. Movements of whole cells can be roughly divided into two functional categories: swimming, or movement through liquid water; and crawling, or movement across a rigid surface.

The physical problems faced by cells attempting to swim have been eloquently summarized by Purcell as 'life at low Reynolds number'. Because viscous forces are many orders of magnitude greater than inertial forces at the speeds, viscosities, and length scales experienced by swimming cells, simple reciprocal motions cannot produce forward propulsion. Instead, most swimming cells use nonreciprocal motions of cell surface projections to swim. For bacterial cells, the rotation of a helical or corkscrew-shaped flagellum has been particularly well-characterized from biological and physical perspectives.

The flagellum is a long filament constructed by the noncovalent polymerization of hundreds of identical protein subunits, called flagellin. The speeds of flagellar swimming range from about 10 to 100 micrometers per second. Some bacteria such as Vibrio cholerae, the causative agent of cholera, have a single flagellum at one pole and swim rapidly. Others, including the common laboratory organism Escherichia coli, have multiple flagella distributed around their surfaces that gather together in a bundle during swimming and can fly apart in a regulated way when the bacterium chooses to change direction.

Rotary flagella have never been found in eukaryotes. Most swimming eukaryotic cells, ranging from human sperm to paramecia, use a second strategy, the propagation of a waveform down a flexible oar. When the cell surface projections used as oars are short and numerous, as on a paramecium, they are called cilia, and when they are long and few, as on sperm, they are called flagella. It is an unfortunate historical accident of terminology that bacterial flagella and eukaryotic flagella share the same name, as they propagates from the front of the cell to the back, a process called metaboly.

The rotary corkscrews and flexible oars described by Purcell are not the only cellular strategies for swimming. Some species of Euglena and related protozoa (unicellular eukaryotes) swim by gradually changing the contour of their surface. This elegant movement, called metaboly, is the low Reynolds number equivalent of the high Reynolds number ice-skater increasing her rotational speed by pulling arm mass inward. In metaboly, the reshifting of cellular contents to generate a local increase in drag that propagates from the front of the cell to the back is sufficient to pull the cell's center of mass forward through the viscous aqueous environment.

The Eutreptiella cell shown herein uses flagella and metaboly for movement, and some related species exhibit phototaxis (movement toward a source of light). Other protozoa offer similarly fascinating motility phenomena that await biophysical investigation. Movement of cells across rigid surfaces can be achieved by an even greater variety of mechanisms than can swimming. The best characterized is amoeboid motility, or crawling motility, a general process shared by eukaryotic cells ranging from soil amoebae to human white blood cells.

In amoeboid motility, a cell attached to a rigid substrate extends forward a projection at its leading edge that then attaches to the substrate. Long thin projections are called filopodia; flat veil-shaped projections, lamellipodia; and thick knobby projections, pseudopodia. All three types of projections are filled with assemblies of cytoskeletal actin filaments. After protrusion and attachment, the crawling cell then contracts to move the cell body forward, and movement continues as a treadmilling cycle of front protrusion and rear retraction. The speed of amoeboid movement can range from less than one micrometer per hour to more than one micrometer per second, depending on the cell type and its degree of stimulation.

Although amoeboid motility is usually accompanied by significant cell shape changes, other forms of surface-based movement are not. The term gliding is used to describe several kinds of movement in which cells slide across a rigid substrate. For some bacteria, gliding appears to be driven by a low Reynolds number analogue of jet propulsion, in which a sticky and cohesive slime is extruded backwards to push the cell forward. Many eukaryotic parasites also perform a kind of gliding, but the mechanism is entirely different, relying on a complex dynamic organization of the pellicle, or skin of the cell.

Twitching motility is a special form of gliding seen only in groups of bacteria. Twitching bacteria throw out long pili, cell surface projections that resemble bacterial flagella except that they are straight instead of helical, that adhere to other nearby bacteria. The bacteria then retract the pili, pulling themselves toward their neighbors. In this manner, colonies of related bacteria growing on surfaces can organize themselves into a multicellular structure. The predatory bacterium Myxococcus xanthus prefers to hunt in packs and uses twitching motility to organize its hunting parties. Lonely Myxococcus switch to a slime-extruding form of gliding.

In addition to movements of the whole cell, as when a paramecium swims after its prey in a drop of pond water or a white blood cell in the human immune system tracks down an invading bacterium, important movements take place within the cell boundaries. One vivid example of this is the process of mitosis, in which the duplicated chromosomes in a dividing eukaryotic cell line up and are materials near a phase transition than gels. Simplifications offered by continuum models help to make complex mechanical processes tractable, with fewer and often more easily measured variables, and models that accurately predict cell behavior then pose the question: how is the continuum description explained in molecular terms?

One of the most important features of cellmotility models, whethermolecular or nonmolecular, is that they make testable, quantitative predictions that allow specific mechanisms to be ruled in or out by subsequent experimental investigation. For example, the original Brownian ratchet model of actin motility was based on diffusion of the cargo and consequently predicted that larger cargo should move more slowly. Experiments in reconstituted systems with polystyrene beads over a range of sizes did not show the expected size dependence, indicating that other factors must dominate and leading to models with improved predictive power.

Predictions from recent models of actin-based motility now await testing. Branching structure models of actin polymerization based on autocatalytic and nucleation branching structures predict differences in actin filament density and force-velocity curves, depending on the exact nature of the biochemical interactions among neighboring actin filaments. A model of actin polymerization in the leading edge predicts a relationship between the number of active filaments and the membrane velocity with a local maximum. With improvements in optical and force microscopy techniques, repeated rounds of collaboration between experimental measurement and theoretical modeling will approach descriptions of cell motility that capture macroscopic behavior and microscopic detail.

Biological cells-inhomogeneous, dynamic, and complex-are unlike any other material familiar to most physical scientists. The challenge of developing a complete physical description of how and why cells move has opened a wide range of experimental and modeling opportunities and given rise to numerous collaborative investigations. The identity and role of many parts of the motility machinery remain to be found, particularly those involved in control and regulation.

New ways of describing material properties of cells that capture the temporal changes and environmental stimulus are needed. New experimental techniques are required to extract physical information from motile cells, and new models are needed to link molecular- and nonmolecular-scale models that can predict newtargets for measurement. Fruitful investigation in cell motility at the interface of biology and physics will demand an increase in the number of scientists fluent in cross-disciplinary languages, biologists who can frame questions in terms ofMonte Carlo simulations and piconewton force measurements, as well as physicists comfortable with genetic analysis and protein purification. In addition to explaining fundamental biological processes that are necessary for survival, investigation of the mechanics of cell motility promises to exemplify interesting dynamic physical properties of living materials and provide tools for nanotechnology and bio-engineering.

CELL ADHESION

Cell adhesion encompasses both cell-cell and cell-matrix interactions and it is the basis upon which tissues and organs are structurally maintained. Cell adhesion proteins usually share three key features:

1) a large extracellular domain that binds to ligands;
2) a single-helix transmembrane domain; and
3) a small cytoplasmic domain that uses adaptor molecules to connect to the cytoskeleton.

A deficiency in adhesion leads to metastatic cancer and apoptotic events, giving the phenomenon considerable clinical relevance. Focal contacts are one class of cell matrix adhesion structure. They occur in migrating cells, such as fibroblasts, and are transient in nature. These cell-matrix interactions are mediated by transmembrane protein receptors called "integrins." On the extracellular surface, these receptors bind to matrix ligands that help anchor cells and promote adhesion. Integrin transmembrane domains connect the extracellular receptor to the intracellular domain, which links integrins to the cytoskeleton.

Sites of cell-matrix contacts in non-migratory cells, such as epithelial cells, are called hemidesmosomes. Integrins provide the primary adhesive contacts in hemidesmosomes while analogous molecules known as cadherins mediate homotypic adhesions between cells. Cadherins are found at the surface of stationary cells and are functionally quite similar to integrins. They also have a large extracellular domain, a single transmembrane helix and a small cytoplamic domain. Inside the cell, various adaptor molecules provide the connections between the external receptor and the cytoskeleton. Cell-cell contacts within specific tissue types are known as adherens junctions. Like hemidesmosomes, in the case of the integrins, adherens junctions are complicated adhesive assemblies with little known about their 3-D structure.

The structural investigation of cell adhesion complexes using electron microscopy. The goal was to produce an in vitro system to facilitate structural studies of the extracellular and intracellular domains in complex with their interacting proteins. For the extracellular domains, the goal was to produce complexes with their specific ligand. For the intracellular domains, the goal was to produce complexes with the adaptor proteins that bind actin filaments. Furthermore, to produce ordered specimens of these adhesive assemblies, suitable for transmission electron microscopy (EM) and three-dimensional image reconstruction, principally electron crystallography.

Electron crystallography has developed as an essential EM technique for obtaining 3-D images of protein arrays. Like all crystallography methods, its major bottle-neck is the production of crystals, of which 2-D arrays are the most useful. A particularly useful method for 2-D crystallization of watersoluble proteins is the lipid monolayer technique first developed by Uzgiris and Kornberg. The fluidity of the different types of lipid monolayer allows for the spontaneous rearrangement of macromolecules at the air-water interface and the possibility of achieving a regular lattice and 2-D crystals.

This monolayer technique both to crystallize proteins and to mimic the assembly of the extracellular and cytoplasmic components of cell adhesion complexes. Images of these arrays were collected using Transmission Electron Microscopy on negatively stained and unstained, frozen-hydrated specimens. Image processing of 2-D crystals yielded 2-D and 3-D information of protein complexes initially involved in focal adhesion assembly. These images were then used to extract atomic detail information by docking high resolution X-ray crystal structures into a molecular envelope. This process can yield information on protein interactions at a resolution of 3 - 4 times that of the actual reconstruction.

This combination of in vitro assembly, electron microscopy and 3-D atomic model building establishes a technique for constructing complexes of adhesive proteins whose spatial relationships are at the present time virtually unknown.

Normal cells obligately adhere to extracellular matrix components, such as fibronectin or collagen for proper control of proliferation. This suggests that the extracellular matrix generates signals for the progression of cells through the cell cycle. Integrins coordinate these signaling pathways from the extracellular domain to the cytoskeleton in focal adhesions.

(spontaneous mutation), especially the process of oxidative deamination;

— exogenous damage caused by external agents such as ultraviolet [UV 200-300nm] radiation from the sun other radiation frequencies, including x-rays and gamma rays, hydrolysis or thermal disruption, certain plant toxins, human-made mutagenic chemicals, especially aromatic compounds that act as DNA intercalating agents cancer chemotherapy and radiotherapy

The replication of damaged DNA before cell division can lead to the incorporation of wrong bases opposite damaged ones. Daughter cells that inherit these wrong bases carry mutations from which the original DNA sequence is unrecoverable (except in the rare case of a back mutation, for example, through gene conversion).

Types of damage

There are four main types of damage to DNA due to endogenous cellular processes:

— *oxidation* of bases [e.g. 8-oxo-7,8-dihydroguanine (8-oxoG)] and generation of DNA strand interruptions from reactive oxygen species,

— *alkylation* of bases (usually methylation), such as formation of 7-methylguanine, 1-methyladenine, O6 methylguanine

— *hydrolysis* of bases, such as deamination, depurination and depyrimidination.

— *mismatch* of bases, due to errors in DNA replication, in which the wrong DNA base is stitched into place in a newly forming DNA strand, or a DNA base is skipped over or mistakenly inserted.

Damage caused by exogenous agents comes in many forms. Some examples are:

— *UV-B light* causes crosslinking between adjacent cytosine and thymine bases creating *pyrimidine dimers*. This is called direct DNA damage.

— *UV-A light* creates mostly free radicals - especially if sunscreen penetrated into the skin. The damage caused by free radicals is called indirect DNA damage.

— *Ionizing radiation* such as that created by radioactive decay or in *cosmic rays* causes breaks in DNA strands.

— *Thermal disruption* at elevated temperature increases the rate of depurination (loss of purine bases from the DNA backbone) and single strand breaks. For example, hydrolytic depurination is seen in the thermophilic bacteria, which grow in hot springs at 85–250 °C. The rate of depurination (300 purine residues per genome per generation) is too high in these species to be repaired by normal repair machinery, hence a possibility of an adaptive response cannot be ruled out.

— *Industrial chemicals* such as vinyl chloride and hydrogen peroxide, and environmental chemicals such as polycyclic hydrocarbons found in smoke, soot and tar create a huge diversity of DNA adducts- ethenobases, oxidized bases, alkylated phosphotriesters and Crosslinking of DNA just to name a few.

UV damage, alkylation/methylation, X-ray damage and oxidative damage are examples of induced damage. Spontaneous damage can include the loss of a base, deamination, sugar ring puckering and tautomeric shift.

Nuclear versus Mitochondrial DNA Damage

In human cells, and eukaryotic cells in general, DNA is found in two cellular locations - inside the nucleus and inside the mitochondria. Nuclear DNA (nDNA) exists as chromatin during non-replicative stages of the cell cycle and is condensed into aggregate structures known as chromosomes during cell division. In either state the DNA is highly compacted and wound up around bead-like proteins called histones. Whenever a cell needs to express the genetic information encoded in its nDNA the required chromosomal region is unravelled, genes located therein are expressed, and then the region is condensed back to its resting conformation.

Mitochondrial DNA (mtDNA) is located inside mitochondria organelles, exists in multiple copies, and is also tightly associated with a number of proteins to form a complex known as the nucleoid. Inside mitochondria, reactive oxygen species (ROS), or free radicals, byproducts of the constant production of adenosine triphosphate (ATP) via oxidative phosphorylation, create a highly oxidative environment that is known to damage mtDNA. A critical enzyme in counteracting the toxicity of these species is superoxide dismutase, which is present in both the mitochondria and cytoplasm of eukaryotic cells.

Senescence and Apoptosis

Senescence, an irreversible state in which the cell no longer divides (mitosis), is a protective response to the shortening of the chromosome ends (telomeres). The telomeres are long regions of repetitive noncoding DNA that cap chromosomes and undergo partial degradation each time a cell undergoes division. In contrast, quiescence is a reversible state of cellular dormancy that is unrelated to genome damage.

Senescence in cells may serve as a functional alternative to apoptosis in cases where the physical presence of a cell for spatial reasons is required by the organism, which serves as a "last resort" mechanism to prevent a cell with damaged DNA from replicating inappropriately in the absence of pro-growth cellular signaling. Unregulated cell division can lead to the formation of a tumor, which is potentially lethal to an organism. Therefore the induction of senescence and apoptosis is considered to be part of a strategy of protection against cancer.

DNA Repair Mechanisms

Cells cannot function if DNA damage corrupts the integrity and accessibility of essential information in the genome (but cells remain superficially functional when so-called "non-essential" genes are missing or damaged). Depending on the type of damage inflicted on the DNA's double helical structure, a variety of repair strategies restore lost information. If possible, cells use the unmodified complementary strand of the DNA or the sister chromatid as a template to losslessly recover the original information. Without access to a template, cells use an error-prone recovery mechanism known as translesion synthesis as a last resort.

Damage to DNA alters the spatial configuration of the helix and such alterations can be detected by the cell. Once damage is localized, specific DNA repair molecules bind at or near the site of damage, inducing other molecules to bind and form a complex that enables the actual repair to take place. The types of molecules involved and the mechanism of repair that is mobilized depend on the type of damage that has occurred and the phase of the cell cycle that the cell is in.

Direct Reversal

Cells are known to eliminate three types of damage to their DNA by chemically reversing it. These mechanisms do not require a template, since the types of damage they counteract can only occur in one of the four bases. Such direct reversal mechanisms are specific to the type of damage incurred and do not involve breakage of the phosphodiester backbone. The formation of thymine dimers (a common type of cyclobutyl dimer) upon

The development of optical and force microscopy techniques over the last several decades has significantly advanced our quantitative understanding of cellmotility at the level of single molecules, collections of molecules, and whole cells. Time-lapse video microscopy produced early measurements of cell movements and internal dynamics. Imaging methods based on polarization, phase, differential interference contrast, and dark-field microscopy can be used with conventional or high-speed strobe light illumination to capture motion of single flagella on sperm, migration of epithelial cells, and contraction of muscle cells.

The development of fluorescent markers capable of localizing specific proteins in time and space has spawned powerful quantitative microscopy techniques. For example, a ratiometric version of fluorescence localization after photobleaching (FLAP) was recently used to show that monomeric actin delivery to the leading edge of motile cells exceeds the speed expected for simple diffusion. At the single molecule level, centroid tracking of a fluorescent marker attached to one head of myosin V, a two-headed processive molecular motor that moves along actin filaments, has revealed that the motor steps are consistent with a hand-over-hand model for walking rather than an inchworm model in which one head always leads the other.

Optical microscopy, and in particular fluorescence microscopy techniques for labeling specific proteins, continues to be a basic tool for quantifying the spatial distribution and time variation of molecular components and cell morphology. While position information guides the cell motility story line, mechanical information is required for a full Newtonian description of the process. Force measurements can be extracted from optical images by allowing motile cells or force-exerting molecules to interact with materials of known mechanical properties.

A dramatic example from more than 20 years ago was the demonstration that fibroblasts cultured on a thin silicone rubber sheet would exert forces sufficient to wrinkle the substrate. Shear force estimates of 0.001 dynes per micron were made by inducing similar wrinkles with a calibrated glass microneedle. Glass microneedles with calibrated spring constants have been used in other experiments to directly measure forces, as in early work investigating muscle contraction.

Recently, glass microneedles and flexible substrates were used to demonstrate the role of tensile stress in microtubule growth and cell motility. Other materials with known mechanical properties can be used to infer force information. For example, estimates of actin polymerization forces on artificial cargo have been obtained from deformation of lipid vesicles. Investigations of the mechanics of single molecules in vitro have motivated the development of optical, magnetic, and atomic force microscopy methods for probing biological forces ranging from 10-12 to 10-6 newtons.

Optical traps or tweezers use focused light to elastically confine a bead near the focus, with trap stiffness depending on optical power. Displacements of a bead linked to a motor protein or other force-generating molecule are converted into force measurements and have been used to observe single steps of the linear stepper motors kinesin and myosin. In magnetic tweezers, a magnetic particle is manipulated by an external magnetic field to generate force. Superparamagnetic beads attached to a macrophage were used to guide cellular extensions with a force as small as 0.5 nanonewtons.

Atomic force microscopy (AFM) has also been used to investigate the mechanical properties of motile cells. Atomic force microscopy is based on the elastic deflection of micromachined cantilevers to measure and apply forces. Originally developed as a surface imaging tool, AFM has found applications in biology such as unfolding single proteins and measuring whole-cell mechanics. Maps of a cell's elastic modulus can be obtained by applying a known force and measuring actual deflection, a technique known as force mapping, which has been applied to quantify mechanical differences in the active and stable edges of motile fibroblasts. On their own, the physical properties revealed by a wide range of measurement techniques provide a glimpse of the mechanical diversity of the single molecules and whole cells. Together, these properties provide inputs for mathematical models designed to capture the macroscopic behavior of cell motility.

A complete understanding of cell motility, involving the coordinated activities of thousands of proteins and signaling molecules, is impossible to capture in a single force measurement or mechanical property. Mathematical modeling provides a link between the observed behavior of cells (speed, shape change) and the properties of single molecules ormodules of constituent proteins. The reductionist strategy of determining the identity and rate constants of, for example, the key molecular players in lamellipodial extension has been largely successful, but modeling must rebuild the process and verify our understanding with predictive power.

Current models can be divided into two categories based on the level of approximation: (1) molecular and (2) nonmolecular or continuum models. Molecular or stochastic models attempt to capture the effect of Brownian motion on the activity of individual proteins and provide a mechanistic explanation of protein behavior. In contrast, nonmolecular or continuum models of cell motility describe movements in terms of bulk or averaged material properties of the cell and its environment.

In the case of Listeria monocytogenes, a bacterial pathogen that harnesses the assembly and disassembly motor of actin polymerization to power its intracellular and intercellular movement, molecular and nonmolecular models have been proposed to describe its motion. The

Brownian ratchet models for actin polymerization, like those of molecular motor movement, relate movement of the bacterium to stochastic fluctuations in the filaments as a function of external load. Inclusion of elastic filament bending and filament attachment properties into the brownian ratchet model yielded a force-velocity curve that shows an exponential decay and is consistent with some observations of Listeria motion.

Continuum models ignore local molecular properties in favor of bulk mechanical properties. In a model by Gerbal et al, Listeria motility is described by elastic energy storage in a compressible gel. Bulk mechanical properties of the gel substitute for the complex network of actin filaments and accessory proteins responsible for bacterialmovement and simplify analysis. Although gel properties used in the model do not account for molecular kinetics, it is consistent with experiments showing 'hopping' behavior of Listeria under some conditions.

Other continuum models have focused on descriptions of the bulk properties of static cells. While early descriptions of the cell as a bag of viscoelastic material have been abandoned, recent models have succeeded in identifying several important material properties of various cells grown under laboratory conditions. The tensegrity analogy for cell structure proposes that the cell can be described as an interconnected array of tension and compression elements, like the architectural tensegrity constructs of Buckminster Fuller. Investigations of prestressed cells using traction-force microscopy have shown a linear increase in shear modulus, consistent with some forms of a tensegrity model.

In contrast, microrheological measurements of smooth muscle cells using a magnetic bead assay have shown that the cells are well-described by a frequency scaling law over at least five orders of magnitude, suggesting that bulk properties of the cell more closely resemble soft glassy simultaneously segregated so that exactly one copy of each chromosome is delivered to each of the two daughter cells. A large, complex cellular structure called the mitotic spindle assembles and then segregates all of the chromosomes simultaneously.

Because of the necessity of accurate chromosome segregation to the life of all cells and the many ways in which this process can go wrong in the development of cancer, chromosome segregation has been a particularly well studied instance of intracellular biological motility. Besides the chromosomes, many other large objects within the cell must be moved quickly to particular locations by directed intracellular transport. The need for directed transport rather than reliance on simple diffusion and trapping to localize cellular components is best imagined in large cells, such as neurons.

The longest neuron in the human body has a single threadlike projection (the axon), a few micrometers in diameter, that reaches from the base of the spine to the foot, a distance of up to one meter.

Most cellular components, including large organelles such as mitochondria, are synthesized within the cell body that lies in the spinal cord and must be delivered down the axon to the synapse, where the neuron forms an electrically active connection with the muscles that flex the toe. Generous estimates of the time it would take for a mitochondrion to diffuse that distance given the cell geometry range from 10 to 100 years. In fact, the mitochondrion can make the journey in just a few days, thanks to intracellular directed transport. Organelle transport is accomplished by a wide variety of mechanisms depending on the specific organelle involved.

The fastest known type of organelle transport is the racetrack-like circulation of chloroplasts in green algae such as Chara, which can achieve speeds greater than 60 micrometers per second.

A final important class of cell motility events comprises rapid and extensive cell shape changes. The best studied of these is skeletal muscle contraction, which has fascinated biologists and physicists for more than 200 years, since Galvani showed that electrical stimulation could cause the muscles of a frog's severed leg to contract.

The proposal in 1953 by Huxley that contraction may be due to the sliding of paired filament arrays relative to each other has proved to be largely correct, and most aspects of muscle stimulation and contraction are now understood in molecular detail. This same type of filament sliding, using closely related molecular components, is responsible for the process of cytokinesis, when two daughter cells are pinched apart after chromosomes are segregated during cell division.

Organisms that lack skeletons and muscles may still exhibit rapid contraction, albeit by a different mechanism. A particularly striking case is the avoidance response of the stalked ciliate Vorticella, as originally described by van Leewenhoek. When this timid tulip-shaped protozoan is touched on its belly, the stalk rapidly coils into a tight spring. With speeds approaching eight centimeters per second and force generation in the range of tens of nanonewtons, the contractile spasmoneme structure within the stalk is among the fastest and most powerful engines known in cell motility.

An opposite type of cellular shape change, rapid extension of a long stiff protrusion, is a common behavior of sperm cells that have contacted the jelly coat of a compatible egg. The rapid extension of this cell surface protrusion through the jelly coat is necessary to allow sperm and egg membrane fusion and proceeds by strikingly different mechanisms in different animal species. A more modest form of the same phenomenon is seen in the mating of the unicellular alga Chlamydomonas.

MOLECULAR MOTORS

Cells generally store chemical energy in two forms: highenergy chemical bonds, such as the phosphoanhydride bonds inATP (adenosine triphosphate); and asymmetric ion gradients across membranes, such as the electrical potential seen in nerve cells. These sources of chemical energy drive all cell processes, from metabolism through DNA replication. The subset of cell proteins and macromolecular complexes that convert chemical energy into mechanical force are generally called molecular motors. Their astonishing variety reflects the diversity of cell movements necessary to life.

Known biological molecular motors may be divided into five general groups:

(1) rotary motors,

(2) linear stepper motors,

(3) assembly and disassembly motors,

(4) extrusion nozzles, and

(5) prestressed springs.

All of the various cell movements already described are performed by ensembles of molecular motors that fall into these categories. Coordinated actions of many small individual components can give rise to large-scale (cellular or organismal) movements. Because the molecular motor appears to be the fundamental unit of biological motility, much experimental and theoretical effort has focused on understanding these motor elements. The best characterized motor in the bacterial kingdom is the tiny rotary motor that enables bacteria to swim. This motor uses ion flux down an electrochemical gradient to drive the rotation of the long, thin helical flagellum at a frequency of about 100 hertz.

In Escherichia coli and its close relatives such as Salmonella typhimurium (a common cause of food poisoning), the motor can rotate in both directions, and switching from counterclockwise to clockwise rotation is under the control of an elaborate regulatory apparatus in the bacterium that detects whether it is moving toward or away from a desirable nutrient source. Some other bacteria spin their flagellar motors in only one direction, but modulate the speed or the frequency of stops. All known biological rotary motors use energy stored in an ion gradient to produce torque.

Most use the gradient of hydrogen ions that is found across the membranes of essentially all living cells; biologists commonly call hydrogen ions protons, and this electrochemical gradient is termed the protonmotive force. In some bacteria, though, such as Vibrio cholerae, sodium ion flux is used to power the flagellar rotary motor. A particularly interesting case of a rotary motor is the F1 ATPase used to synthesize ATP in mitochondria. This molecular motor (or dynamo) converts chemical energy in the form of a proton gradient into mechanical energy in the form of a unidirectional rotation and then back again into chemical energy in the form of ATP. In the presence of excess ATP, the motor can be made to run backwards, with no apparent loss of efficiency, pumping protons against their electrochemical gradient at the expense of ATP hydrolysis.

Linear stepper motors are much more common in eukaryotic forms of motility. These motors move along preassembled linear tracks by coupling binding to the track with ATP hydrolysis and a large-scale protein conformational change. The first linear stepper motor to be purified and characterized was myosin, the motor that drives filament sliding in skeletal muscle contraction. The track for myosin is the actin filament, or microfilament, a helical polymer formed by noncovalent self-association of identical globular subunits. It is now clear that skeletal

Analogously, cadherins mediate a similar adhesive phenomenon that exists at sites of cell-cell contact. Both integrin and cadherin adhesive components provide a scaffolding apparatus that structurally supports the cell while activating a signaling cascade to control growth and development. Signals from adhesive receptors are integrated with those from growth factors to organize the cytoskeleton and stimulate kinase cascades for regulating early gene expression. Multiple adhesive dependent pathways are involved in preventing apoptosis, stopping migration and facilitating differentiation. Ultimately, it is the combined extracellular matrix-integrin signaling efforts that dictate whether a cell will survive, proliferate, or exit the cell cycle. Determining the structural aspects of such integrated systems would provide key insight as to how these events occur.

Intracellular and Extracellular Scaffolding

α5β1-Integrin

Integrins are heterodimeric proteins composed of two subunits; one α-and one β- polypeptide chain. Molecular biology has identified 16 α- and 8 β- subunits but only 22 of the 128 possible α-, bcombinations exist. At the start of this investigation, the most extensively studied integrin was the α5 β1-integrin, which is a fibronectin receptor. This seemed like an ideal candidate to use for building assemblies, since some information was known of its sequence and physical nature. Both the α- and β-subunits of the α5β1-integrin have a small cytoplasmic domain. The asubunit acts as a regulatory domain while the β- subunit binds to the cytoskeleton.

In the absence of ligand binding to the extracellular domain, the α5-integrin cytoplasmic tail is thought to sterically block the cytoskeletal binding sites of the b-subunit. Following fibronectin binding to the extracellular receptor, the cytoplasmic α5-domain dissociates from the β-subunit, leaving the latter free to interact with adaptor molecules that bind to the cytoskeleton. In the case of migratory cells, the β1-integrins interact with both of the structural, actin cross-linkers: talin and αactinin. This is the β-integrin used in this experiments.

The N-terminal, extracellular domain is colored purple, while the transmembraneous region is depicted green. The cytoplasmic sequence is labeled to indicate both the high and low affinity α-actinin binding sites and the talin binding site that overlaps them. The cytoplasmic domain of the β1-integrin consists of 41 amino acids. This domain has binding sites for talin and two alternative sites that bind to α-actinin. The protein vinculin also has a supporting role in integrin adhesion. It does not bind directly to integrins, but it has binding sites for talin, F-actin and α-actinin.

It may therefore, help anchor these proteins to the integrin complex. This experiment emphasized Cytoplasmic linkers—talin, α–actinin, vinculin and actin.

Talin

Much of what is known of the overall structure of talin comes from the work of Moloney et al, 1987. Talin (270 kDa) is an elongated flexible actin cross-linking protein that is thought to be 56-60 nm in length. It migrates anomalously on SDS-PAGE gels at approximately 230kDa. The monomeric form of talin dominates at protein concentrations <0.72mg/ml. Above this concentration it self-associates to form antiparallel homodimers. In low ionic strength (5-20 mM NaCl), it has a Stokes radius of 6.5 nm and a sedimentation coefficient of 9.4, suggesting that it resembles an asymmetric globular mass.

At approximately 200 mM NaCl, it is thought be more elongated with a Stokes radius of 7.7 nm and a sedimentation coefficient of 8.8. Under physiological conditions (150 mM NaCl) it has the potential to simultaneously bind to the β1-integrin and cross-link actin filaments. The overall structure of talin consists of an N-terminal head domain, followed by a helical, rodlike C-terminal tail domain. The 47kDa N-terminal head contains both b-integrin and actin binding sites within a FERM (Band 4.1-Ezrin-Radixin-Moesin) homology domain.

The FERM domain is linked by a calpain protease sensitive region of random coil to the 190 kDa rod domain within the C-terminus of the molecule. The rod-domain is 79% ahelical by CD spectroscopy and contains three vinculin binding sites and two additional actin-binding domains. The vinculin sites are non-overlapping repeats of short amphipathic helices, as is similarly found in the adhesion protein paxillin.

α-Actinin

α–Actinin (105 kDa), a member of the spectrin superfamily, is a structural protein that is found throughout the cytoplasm but is concentrated at focal adhesions, Z-disks and homologous structures. It is traditionally known to form bipolar crosslinks between actin filaments of the Z-disk of skeletal muscle. Alternatively, in smooth and non-muscle cells, α–actinin is also known to anchor polar actin filaments to the cell membrane.

It has a similar role as talin in linking the cell membrane to the cytoskeleton, but was originally thought to be insufficient on its own in recruiting actin filaments to clustered integrins. Its role in cell attachment is still somewhat obscure. It is often undetected in focal adhesions by immunofluorscence or EM, even though it is prevalent in stress fibers. It was later suggested by Pavalko et al that the dense structure of focal adhesions render antibodies inaccessible to the α–actinin present.

Within the cell, α–actinin exists as a 36 nm antiparallel homodimer with an N-terminal actinbinding region composed of two calponin homology domains, a central rod region, and a C-terminal domain which contains a pair of calmodulin-like EF hand, calcium binding motifs.

The N-terminal actin-binding domain is thought to be in a compact conformation in its functional state. It is connected to the rod domain by a flexible linker region. The central portion of the molecule contains four tandem spectrin-like repeats (3-helix motifs) that are symmetrically arranged around a two-fold axis and have a 90° left-handed twist along the long axis. Another flexible linker connects the end of the fourth spectrin repeat to the calmodulin-like domain. This domain contains a pair of calcium binding motifs that are thought to potentially regulate its ability to bind actin in non-muscle, cytoskeletal isoforms.

Vinculin

Vinculin (120 kDa) is a highly conserved protein that provides a structural and functional linkage between actin and the b1-integrin via talin and alpha-actinin. Vinculin has a 90 kDa "head" region connected to a flexible "tail" domain by a short hinge. The binding sites on vinculin for talin, α–actinin, and Vasodilater Associated Stimulating Protein (VASP) are blocked by a strained head-to-tail interaction. This association is relieved by the membrane localized cofactor phosphytidyl-inositol-diphosphate, PIP_2. PIP_2 activates the tail domain, which anchors it to the cell membrane, leaving its head free to interact with other cytoskeletal components in close proximity.

Recently, a 1.8Å crystal structure of the vinculin tail domain was solved by X-ray crystallography.

Biological implications from this structure are that the vinculin tail domain directly interacts with the cell membrane via PIP_2 and this interaction has an essential supporting role in focal adhesion assembly. It may reversibly associate with the membrane in its active state contributing to the dynamic remodeling of adhesive components.

Epithelial cadherin

Cadherins (150kD) are homotypic, dimeric cell mediators that are primarily responsible for vertebrate cell-cell adhesion. The classical cadherin arrangement allows for contact between dimers on the same surface of the cell as well as opposing cells. This is the basis for the linear zipper model in which both cis-(same cell) or trans-(between cells) arrangements of cadherin lattices are possible. From this zipper model stems the idea that a combination of *cis*- and transinteractions, along with calcium ions, help to physically reinforce cell-cell attachments and produce a more rigid cadherin molecule, which is otherwise somewhat flexible on the cell surface.

Cadherins are localized to adherens junctions and other tight junctions, such as desmosomes, of various solid tissues. At adherens junctions, cadherins interact with catenins, which link them to the actin network. Similarly, in desmosomes, cadherins interact with desmoplakin to attach to intermediate filaments. Cadherins play an extremely important role in embryonic development and tissue formation because they must constantly change as the embryonic epithelium is altered.

The x-ray crystal structure of the ectodomain of the Type I classical cadherin (C-cadherin) from adherens junctions was solved to 3.1 Å. These ectodomains dimerize and cluster in the presence of millimolar calcium concentrations to trigger an intracellular interaction of their cytoplasmic tails with the actin cytoskeleton. The cytoskeletal adaptors α– and β-catenin are required for the actin linkage.

The cadherin adhesive system is quite analogous to that of the integrin adhesive system. The Ecadherin construct allowed us to apply our in vitro assembly technique to also obtain structural information about the extracellular side of adhesive cells. Thus, our monolayer assembly technique developed for the integrin, α–actinin-vinculin complex provides with a tool for structural studies of virtually any type of cell surface that one may wish to examine.

REFERENCES

Albrecht-Buehler, G., "Does the geomentric design of centrioles imply their function?", *Cell Motility*, 1981

______________ ., "Is Cytoplasm Intelligent too?", In: *Muscle and Cell Motility* VI (ed. J. Shay), 1985.

______________ ., "The cellular infrared detector appears to be contained in the centrosome", *Cell Motiltiy and the Cytoskeleton* , 1994.

______________ ., "Changes of cell behavior by near-infrared signals"m *Cell Motiltiy and the Cytoskeleton*, 1995.

7

DNA Repair

DNA repair refers to a collection of processes by which a cell identifies and corrects damage to the DNA molecules that encode its genome. In human cells, both normal metabolic activities and environmental factors such as UV light can cause DNA damage, resulting in as many as 1 million individual molecular lesions per cell per day. Many of these lesions cause structural damage to the DNA molecule and can alter or eliminate the cell's ability to transcribe the gene that the affected DNA encodes. Other lesions induce potentially harmful mutations in the cell's genome, which affect the survival of its daughter cells after it undergoes mitosis. Consequently, the DNA repair process is constantly active as it responds to damage in the DNA structure.

The rate of DNA repair is dependent on many factors, including the cell type, the age of the cell, and the extracellular environment. A cell that has accumulated a large amount of DNA damage, or one that no longer effectively repairs damage incurred to its DNA, can enter one of three possible states:

— an irreversible state of dormancy, known as senescence

— cell suicide, also known as apoptosis or programmed cell death

— unregulated cell division, which can lead to the formation of a tumor that is cancerous

The DNA repair ability of a cell is vital to the integrity of its genome and thus to its normal functioning and that of the organism. Many genes that were initially shown to influence lifespan have turned out to be involved in DNA damage repair and protection.[2] Failure to correct molecular lesions in cells that form gametes can introduce mutations into the genomes of the offspring and thus influence the rate of evolution.

DNA DAMAGE

DNA damage, due to environmental factors and normal metabolic processes inside the cell, occurs at a rate of 1,000 to 1,000,000 molecular lesions per cell per day. While this constitutes only 0.000165% of the human genome's approximately 6 billion bases (3 billion base pairs), unrepaired lesions in critical genes (such as tumor suppressor genes) can impede a cell's ability to carry out its function and appreciably increase the likelihood of tumor formation.

The vast majority of DNA damage affects the primary structure of the double helix; that is, the bases themselves are chemically modified. These modifications can in turn disrupt the molecules' regular helical structure by introducing non-native chemical bonds or bulky adducts that do not fit in the standard double helix. Unlike proteins and RNA, DNA usually lacks tertiary structure and therefore damage or disturbance does not occur at that level. DNA is, however, supercoiled and wound around "packaging" proteins called histones (in eukaryotes), and both superstructures are vulnerable to the effects of DNA damage.

Sources of Damage

DNA damage can be subdivided into two main types:

— endogenous damage such as attack by reactive oxygen species produced from normal metabolic byproducts muscle myosin is just one member of a large family of related myosin motor proteins.

All use actin filaments as a track and all share a similar motor 'head' domain, but the 'tail' domain that attaches the motor to its cargo is highly divergent. Some tails cause dimerization, leading to the formation of a two-headed motor, as is the case for muscle myosin (myosin II) and several other classes. Currently, there are at least 18 different classes of myosins known, and each class may comprise dozens of different members even in a single organism.

Various forms of myosin in humans are responsible for biological movements as diverse as muscle contraction, cell division, pigment granule transport in the skin, and sound adaptation in the hair cells of the inner ear. The exact mechanical features of each type of myosin motor head, including the stepping speed, stepping distance, fraction of time bound to the track versus free, and amount of force generated, appear to be carefully tuned to their biological functions. Myosins of different classes have step sizes ranging from 5 to 36 nanometers and generate forces generally on the order of a few piconewtons, performing between 1 and 20 steps per second.

In the case of myosin V, which is responsible for transport of pigment granules in mammalian skin cells, a single molecule of the motor protein may be sufficient for granulemovement, while in skeletal muscle, nearly crystalline arrays of hundreds of myosin II molecules act in concert to drive the shortening of each sarcomere. A unique and speedoptimized plant myosin drives the extremely rapid circulation of chloroplasts in Chara, while another unusual myosin is associated with gliding movements of eukaryotic parasites. The mechanisms by which different myosins can exhibit such a wide variety of mechanochemical properties are not yet fully understood.

Similar functional tuning of motor mechanical properties is found in another abundant and diverse family of linear stepper motors, the kinesins, which use microtubules rather than actin filaments as a track. Like the myosins, the kinesins are most similar in the motor head domain and show great divergence in the tail. Kinesins are involved in many types of intracellular transport, including transport of organelles down nerve axons and chromosome segregation. Although myosins and kinesins use different tracks and have different mechanisms for coupling the energy derived from ATP hydrolysis to the protein conformational change associated with force generation, recent work determining the atomic structure of canonical members of both families has revealed that the core protein structures are remarkably similar, an almost certain indication that myosins and kinesins are derived from the same common ancestor.

The common motor core represents a fundamental physical unit coupling ATP hydrolysis to a small movement of two alpha helices in the protein structure; additional protein levers and latches convert this small molecular movement into the large and various functional steps of the motor proteins. Another family of linear stepper motors that walks on microtubules, the dyneins, is less well-characterized. It is not yet known whether dyneins also share a common ancestor with kinesin and myosin or whether they might represent an independent, convergent development of stepper motor function. Dyneins also play a part in intracellular transport and chromosome segregation. The fastest and strongest dyneins are those responsible for the sliding and bending of the microtubules inside eukaryotic flagella and cilia during cell swimming.

In addition to serving as the tracks for some kinds of linear stepper motors, microtubules and actin filaments can assemble and disassemble rapidly to change the shape of the cell and to produce force on their own. In these forms of biological force generation, the chemical energy comes from nonequilibrium protein polymerization, although ultimately the cellular pools of polymerizing actin and tubulin subunits are maintained in a steady state far from chemical equilibrium due to a coupling between protein polymerization and ATP or GTP (guanosine triphosphate) hydrolysis. Force generated by actin polymerization is responsible for extension of the acrosome in some kinds of animal sperm, as well as the mating structure in Chlamydomonas.

With the filaments arranged in a slightly different geometry, actin polymerization is also the major driving force for cell protrusion at the leading edge in amoeboid motility and for the movement of certain kinds of bacterial pathogens, including Listeria monocytogenes. In an unusual form of amoeboid motility, the crawling sperm of nematodes (roundworms) uses an assembly and disassembly motor built by a unique protein, MSP (major sperm protein). The striking degree of resemblance between the crawling nematode sperm and a conventional crawling cell that uses actin assembly and disassembly for protrusion at the leading edge demonstrates that the physical rules governing the mechanics of cell motility are largely independent of the exact molecular identity of their components.

Whereas rotary motors and linear stepper motors have been characterized in great physical detail and force measurements have been performed on single molecules of each class, assembly and disassembly motors are poorly understood. Only a few measurements of the amount of force generated by single microtubules have yet been made, and there is to date no direct measurement of the force generated by polymerization of a single actin filament.

The final two classes of biological motors, the extrusion motors and the prestressed springs, demand even more urgent attention. While the biological phenomena are well described and some initial physical and mathematical characterization has begun for each, the physical models currently lack predictive detail. Most interesting and complex forms of biological movement require the coordinated action of many kinds of molecular motors acting simultaneously. For example, the construction of the mitotic spindle and the process of chromosome segregation require forces generated by microtubule assembly and disassembly, as well as the action of about a dozen different kinesin and dynein motors.

Similarly, amoeboid crawling motility requires actin filament assembly and disassembly, as well as at least three kinds of myosin motors. While reductionist techniques have enabled us to identify the molecules involved in many forms of biological movement, a complete understanding of the process demands a complementary synthetic approach to bring together biological characterization with physical measurements and mathematical modeling of properties emergent from large collections of motor elements.

MOTILITY MEASUREMENTS

Identification of the molecular machineries that power cell motility has raised two major questions: (1) what is the mechanical behavior of the individual motor protein components and (2) how are proteins with diverse mechanical roles organized to produce coordinated mesoscopic and macroscopicmovements? Answers to the first question require experimental methods capable of isolating the biophysical activity of individual or small groups of proteins, and answers to the second rely on mathematical models capable of predicting collective behavior from measured properties.

irradiation with UV light results in an abnormal covalent bond between adjacent thymidine bases.

The photoreactivation process directly reverses this damage by the action of the enzyme photolyase, whose activation is obligately dependent on energy absorbed from blue/UV light (300–500nm wavelength) to promote catalysis. Another type of damage, methylation of guanine bases, is directly reversed by the protein methyl guanine methyl transferase (MGMT), the bacterial equivalent of which is called as ogt. This is an expensive process because each MGMT molecule can only be used once; that is, the reaction is stoichiometric rather than catalytic.

A generalized response to methylating agents in bacteria is known as the adaptive response and confers a level of resistance to alkylating agents upon sustained exposure by upregulation of alkylation repair enzymes. The third type of DNA damage reversed by cells is certain methylation of the bases cytosine and adenine.

Single Strand Damage

When only one of the two strands of a double helix has a defect, the other strand can be used as a template to guide the correction of the damaged strand. In order to repair damage to one of the two paired molecules of DNA, there exist a number of excision repair mechanisms that remove the damaged nucleotide and replace it with an undamaged nucleotide complementary to that found in the undamaged DNA strand.

Base excision repair (BER), which repairs damage to a single nucleotide caused by oxidation, alkylation, hydrolysis, or deamination. The base is removed with glycosylase and ultimately replaced by repair synthesis with DNA ligase.

Nucleotide excision repair (NER), which repairs damage affecting longer strands of 2–30 bases. This process recognizes bulky, helix-distorting changes such as thymine dimers as well as single-strand breaks (repaired with enzymes such UvrABC endonuclease). A specialized form of NER known as Transcription-Coupled Repair (TCR) deploys high-priority NER repair enzymes to genes that are being actively transcribed.

Mismatch repair (MMR), which corrects errors of DNA replication and recombination that result in mispaired (but normal, that is non- damaged) nucleotides following DNA replication.

Double-strand Breaks

Double-strand breaks (DSBs), in which both strands in the double helix are severed, are particularly hazardous to the cell because they can lead to genome rearrangements. Two mechanisms exist to repair DSBs: non-homologous end joining (NHEJ) and recombinational repair (also known as template-assisted repair or homologous recombination repair).

In NHEJ, DNA Ligase IV, a specialized DNA Ligase that forms a complex with the cofactor XRCC4, directly joins the two ends. To guide accurate repair, NHEJ relies on short homologous sequences called microhomologies present on the single-stranded tails of the DNA ends to be joined. If these overhangs are compatible, repair is usually accurate.

NHEJ can also introduce mutations during repair. Loss of damaged nucleotides at the break site can lead to deletions, and joining of nonmatching termini forms translocations. NHEJ is especially important before the cell has replicated its DNA, since there is no template available for repair by homologous recombination. There are "backup" NHEJ pathways in higher eukaryotes. Besides its role as a genome caretaker, NHEJ is required for joining hairpin-capped double-strand breaks induced during V(D)J recombination, the process that generates diversity in B-cell and T-cell receptors in the vertebrate immune system.

Recombinational repair requires the presence of an identical or nearly identical sequence to be used as a template for repair of the break. The enzymatic machinery responsible for this repair process is nearly identical to the machinery responsible for chromosomal crossover during meiosis. This pathway allows a damaged chromosome to be repaired using a sister chromatid (available in G2 after DNA replication) or a homologous chromosome as a template. DSBs caused by the replication machinery attempting to synthesize across a single-strand break or unrepaired lesion cause collapse of the replication fork and are typically repaired by recombination.

Topoisomerases introduce both single- and double-strand breaks in the course of changing the DNA's state of supercoiling, which is especially common in regions near an open replication fork. Such breaks are not considered DNA damage because they are a natural intermediate in the topoisomerase biochemical mechanism and are immediately repaired by the enzymes that created them.

Translesion Synthesis

Translesion synthesis is a DNA damage tolerance process that allows the DNA replication machinery to replicate past DNA lesions such as thymine dimers or AP sites. It involves the switching out of regular DNA polymerases for specialized translesion polymerases, often with larger active sites that can facilitate the insertion of bases opposite damaged nucleotides. The polymerase switching is thought to be mediated by, among other factors, the post-translational modification of the replication processivity factor PCNA.

Translesion synthesis polymerases often have low fidelity (high propensity to insert wrong bases) relative to regular polymerases. However, many are extremely efficient at inserting correct bases opposite specific types of damage.

GLOBAL RESPONSE TO DNA DAMAGE

Cells exposed to ionizing radiation, ultraviolet light or chemicals are prone to acquire multiple sites of bulky DNA lesions and double strand breaks. Moreover, DNA damaging agents can damage other biomolecules such as proteins, carbohydrates, lipids and RNA. The accumulation of damage, specifically double strand breaks or adducts stalling the replication forks, are among known stimulation signals for a global response to DNA damage. The global response to damage is an act directed toward the cells' own preservation and triggers multiple pathways of macromolecular repair, lesion bypass, tolerance or apoptosis. The common features of global response are induction of multiple genes, cell cycle arrest, and inhibition of cell division.

DNA Damage Checkpoints

After DNA damage, cell cycle checkpoints are activated. Checkpoint activation pauses the cell cycle and gives the cell time to repair the damage before continuing to divide. DNA damage checkpoints occur at the G1/S and G2/M boundaries. An intra-S checkpoint also exists. Checkpoint activation is controlled by two master kinases, ATM and ATR. ATM responds to DNA double-strand breaks and disruptions in chromatin structure, whereas ATR primarily responds to stalled replication forks. These kinases phosphorylate downstream targets in a signal transduction cascade, eventually leading to cell cycle arrest. A class of checkpoint mediator proteins including BRCA1, MDC1, and 53BP1 has also been identified. These proteins seem to be required for transmitting the checkpoint activation signal to downstream proteins.

p53 is an important downstream target of ATM and ATR, as it is required for inducing apoptosis following DNA damage. At the G1/S checkpoint, p53 functions by deactivating the CDK2/cyclin E complex. Similarly, p21 mediates the G2/M checkpoint by deactivating the CDK1/cyclin B complex.

Prokaryotic SOS Response

The SOS response is the term used to describe changes in gene expression in Escherichia coli and other bacteria in response to DNA damage. The prokaryotic SOS system is regulated by two key proteins: LexA and RecA. The LexA homodimer is a transcriptional repressor that binds to operator sequences commonly referred to as SOS boxes. It is known that LexA regulates transcription of approximately 48 genes including the lexA and recA genes. The most common cellular signals activating the SOS response are regions of single stranded DNA (ssDNA), arising from stalled replication forks or double strand breaks, which are processed by DNA helicase to separate the two DNA strands. Once the DNA damage is repaired or bypassed using polymerases or through recombination, the amount of single-stranded DNA in cells is decreased, lowering the amounts of RecA filaments decreases cleavage activity of LexA homodimer which subsequently binds to the SOS boxes near promoters and restores normal gene expression.

Eukaryotic Transcriptional Responses

Eukaryotic cells exposed to DNA damaging agents also activate important defensive pathways by inducing multiple proteins involved in DNA repair, cell cycle checkpoint control, protein trafficking and degradation. Such genome wide transcriptional response is very complex and tightly regulated, thus allowing coordinated global response to damage. Exposure of yeast Saccharomyces cerevisiae to DNA damaging agents results in overlapping but distinct transcriptional profiles. Similarities to environmental shock response indicates that a general global stress response pathway exist at the level of transcriptional activation.

In contrast, different human cell types respond to damage differently indicating an absence of a common global response. The probable explanation for this difference between yeast and human cells may be in the heterogeneity of mammalian cells. In an animal different types of cells are distributed amongst different organs which have evolved different sensitivities to DNA damage.

In general global response to DNA damage involves expression of multiple genes responsible for postreplication repair, homologous recombination, nucleotide excision repair, DNA damage checkpoint, global transcriptional activation, genes controlling mRNA decay and many others. The vast amount of damage to a cell leaves it with an important decision; undergo apoptosis and die, or survive at the cost of living with a modified genome. An increase in tolerance to damage can lead to an increased rate of survival which will allow a greater accumulation of mutations. Yeast Rev1 and human polymerase are members of [Y family translesion DNA polymerases present during global response to DNA damage and are responsible for enhanced mutagenesis during a global response to DNA damage in eukaryotes.

DNA REPAIR AND AGING

Pathological Effects in DNA Repair

Experimental animals with genetic deficiencies in DNA repair often show decreased lifespan and increased cancer incidence. For example, mice deficient in the dominant NHEJ pathway and in telomere maintenance mechanisms get lymphoma and infections more often, and consequently have shorter lifespans than wild-type mice. Similarly, mice deficient in a key repair and transcription protein that unwinds DNA helices have premature onset of aging-related diseases and consequent shortening of lifespan. However, not every DNA repair deficiency creates exactly the predicted effects; mice deficient in the NER pathway exhibited shortened lifespan without correspondingly higher rates of mutation.

If the rate of DNA damage exceeds the capacity of the cell to repair it, the accumulation of errors can overwhelm the cell and result in early senescence, apoptosis or cancer. Inherited diseases associated with faulty DNA repair functioning result in premature aging, increased sensitivity to carcinogens, and correspondingly increased cancer risk. On the other hand, organisms with enhanced DNA repair systems, such as *Deinococcus radiodurans*, the most radiation-resistant known organism, exhibit remarkable resistance to the double strand break-inducing effects of radioactivity, likely due to enhanced efficiency of DNA repair and especially NHEJ. It is noteworthy that some works suggest that if a DNA damage event occurs during the self repair process then the combination of the two events will exert an effect greater than the sum of the individual events (if they occurred with a long time delay between them), this is the basis of the *second event theory* favoured by C. Busby.

Longevity and Caloric Restriction

A number of individual genes have been identified as influencing variations in lifespan within a population of organisms. The effects of these genes is strongly dependent on the environment, particularly on the organism's diet. Caloric restriction reproducibly results in extended lifespan in a variety of organisms, likely via nutrient sensing pathways and decreased metabolic rate. The molecular mechanisms by which such restriction results in lengthened lifespan are as yet unclear; however, the behavior of many genes known to be involved in DNA repair is altered under conditions of caloric restriction.

For example, increasing the gene dosage of the gene SIR-2, which regulates DNA packaging in the nematode worm *Caenorhabditis elegans*, can significantly extend lifespan. The mammalian homolog of SIR-2 is known to induce downstream DNA repair factors involved in NHEJ, an activity that is especially promoted under conditions of caloric restriction. Caloric restriction has been closely linked to the rate of base excision repair in the nuclear DNA of rodents, although similar effects have not been observed in mitochondrial DNA.

Interestingly, the *C. elegans* gene AGE-1, an upstream effector of DNA repair pathways, confers dramatically extended lifespan under free-feeding conditions but leads to a decrease in reproductive fitness under conditions of caloric restriction. This observation supports the pleiotropy theory of the biological origins of aging, which suggests that genes conferring a large survival advantage early in life will be selected for even if they carry a corresponding disadvantage late in life.

Medicine and DNA Repair Modulation

Hereditary DNA repair disorders

Defects in the NER mechanism are responsible for several genetic disorders, including:

- xeroderma pigmentosum: hypersensitivity to sunlight/UV, resulting in increased skin cancer incidence and premature aging
- Cockayne syndrome: hypersensitivity to UV and chemical agents
- trichothiodystrophy: sensitive skin, brittle hair and nails

Mental retardation often accompanies the latter two disorders, suggesting increased vulnerability of developmental neurons.

Other DNA repair disorders include:

— Werner's syndrome: premature aging and retarded growth
— Bloom's syndrome: sunlight hypersensitivity, high incidence of malignancies (especially leukemias).
— ataxia telangiectasia: sensitivity to ionizing radiation and some chemical agents

All of the above diseases are often called "segmental progerias" ("accelerated aging diseases") because their victims appear elderly and suffer from aging-related diseases at an abnormally young age.

Other diseases associated with reduced DNA repair function include Fanconi's anemia, hereditary breast cancer and hereditary colon cancer.

DNA repair and cancer

Inherited mutations that affect DNA repair genes are strongly associated with high cancer risks in humans. Hereditary nonpolyposis colorectal cancer (HNPCC) is strongly associated with specific mutations in the DNA mismatch repair pathway. BRCA1 and BRCA2, two famous mutations conferring a hugely increased risk of breast cancer on carriers, are both associated with a large number of DNA repair pathways, especially NHEJ and homologous recombination.

Cancer therapy procedures such as chemotherapy and radiotherapy work by overwhelming the capacity of the cell to repair DNA damage, resulting in cell death. Cells that are most rapidly dividing - most typically cancer cells - are preferentially affected. The side effect is that other non-cancerous but rapidly dividing cells such as stem cells in the bone marrow are also affected. Modern cancer treatments attempt to localize the DNA damage to cells and tissues only associated with cancer, either by physical means (concentrating the therapeutic agent in the region of the tumor) or by biochemical means (exploiting a feature unique to cancer cells in the body).

DNA Repair and Evolution

The basic processes of DNA repair are highly conserved among both prokaryotes and eukaryotes and even among bacteriophage (viruses that infect bacteria); however, more complex organisms with more complex genomes have correspondingly more complex repair mechanisms. The ability of a large number of protein structural motifs to catalyze relevant chemical reactions has played a significant role in the elaboration of repair mechanisms during evolution.

The fossil record indicates that single celled life began to proliferate on the planet at some point during the Precambrian period, although exactly when recognizably modern life first emerged is unclear. Nucleic acids became the sole and universal means of encoding genetic information, requiring DNA repair mechanisms that in their basic form have been inherited by all extant life forms from their common ancestor. The emergence of Earth's oxygen-rich atmosphere due to photosynthetic organisms, as well as the presence of potentially damaging free radicals in the cell due to oxidative phosphorylation, necessitated the evolution of DNA repair mechanisms that act specifically to counter the types of damage induced by oxidative stress.

REFERENCES

Friedberg E.C, Walker G.C, Siede W, Wood R.D., Schultz R.A., Ellenberger T., *DNA Repair and Mutagenesis, part 3,* ASM Press. 2nd ed. 2006.

Bakkenist C.J, Kastan M.B., "DNA damage activates ATM through intermolecular autophosphorylation and dimer dissociation", *Nature,* Jan 30 2003.

Wei, Qingyi; Lei Li, David Chen., "DNA Repair, Genetic Instability, and Cancer", *World Scientific,* 2007.

8

Programmed Cell Death

Programmed cell-death (PCD) is death of a cell in any form, mediated by an intracellular program. In contrast to necrosis, which is a form of cell-death that results from acute tissue injury and provokes an inflammatory response, PCD is carried out in a regulated process which generally confers advantage during an organism's life-cycle. PCD serves fundamental functions during both plant and metazoa (multicellular animals) tissue development.

— Apoptosis or Type I cell-death

— Autophagic or Type II cell-death (*cytoplasmic*: characterized by the formation of large vacuoles which eat away organelles in a specific sequence prior to the nucleus being destroyed.)

Besides these two types of PCD, other pathways have been discovered. Called "non-apoptotic programmed cell-death" (or "caspase-independent programmed cell-death" or "necrosis-like programmed cell-death") these alternative routes to death are as efficient as apoptosis and can function as either backup mechanisms or the main type of PCD. Plant cells undergo particular processes of PCD which are similar to autophagic cell death. However, some common features of PCD are highly conserved in both plants and metazoa.

APOPTOSIS

Apoptosis is a form of programmed cell death in multicellular organisms. It is one of the main types of programmed cell death (PCD) and involves a series of biochemical events leading to a characteristic cell morphology and death, in more specific terms, a series of biochemical events that lead to a variety of morphological changes, including blebbing, changes to the cell membrane such as loss of membrane asymmetry and attachment, cell shrinkage, nuclear fragmentation, chromatin condensation, and chromosomal DNA fragmentation (1-4). Processes of disposal of cellular debris whose results do not damage the organism differentiate apoptosis from necrosis.

In contrast to necrosis, which is a form of traumatic cell death that results from acute cellular injury, apoptosis, in general, confers advantages during an organism's life cycle. For example, the differentiation of fingers and toes in a developing human embryo occurs because cells between the fingers apoptose; the result is that the digits are separate. Between 50 billion and 70 billion cells die each day due to apoptosis in the average human adult. For an average child between the ages of 8 and 14, approximately 20 billion to 30 billion cells die a day. In a year, this amounts to the proliferation and subsequent destruction of a mass of cells equal to an individual's body weight.

Research on apoptosis has increased substantially since the early 1990s. In addition to its importance as a biological phenomenon, defective apoptotic processes have been implicated in an extensive variety of diseases. Excessive apoptosis causes hypotrophy, such as in ischemic damage, whereas an insufficient amount results in uncontrolled cell proliferation, such as cancer.

FUNCTIONS

Cell termination

Apoptosis can occur when a cell is damaged beyond repair,

infected with a virus, or undergoing stress conditions such as starvation. DNA damage from ionizing radiation or toxic chemicals can also induce apoptosis via the actions of the tumour-suppressing gene *p53*. The "decision" for apoptosis can come from the cell itself, from the surrounding tissue, or from a cell that is part of the immune system. In these cases apoptosis functions to remove the damaged cell, preventing it from sapping further nutrients from the organism, or to prevent the spread of viral infection.

Apoptosis also plays a role in preventing cancer; if a cell is unable to undergo apoptosis, due to mutation or biochemical inhibition, it can continue dividing and develop into a tumour. For example, infection by papillomaviruses causes a viral gene to interfere with the cell's p53 protein, an important member of the apoptotic pathway. This interference in the apoptotic capability of the cell plays a critical role in the development of cervical cancer.

Homeostasis

In the adult organism, the number of cells is kept relatively constant through cell death and division. Cells must be replaced when they become diseased or malfunctioning; but proliferation must be compensated by cell death. This balancing process is part of the homeostasis required by living organisms to maintain their internal states within certain limits. Some scientists have suggested *homeodynamics* as a more accurate term. The related term *allostasis* reflects a balance of a more complex nature by the body.

Homeostasis is achieved when the rate of mitosis (cell division) in the tissue is balanced by cell death. If this equilibrium is disturbed, one of two potentially fatal disorders occurs:

— The cells are dividing faster than they die, effectively developing a tumor.

— The cells are dividing slower than they die, which results in a disorder of cell loss.

The organism must orchestrate a complex series of controls to keep homeostasis tightly controlled, a process that is ongoing for the life of the organism and involves many different types of cell signaling. Impairment of any one of these controls can lead to a diseased state; for example, dysregulation of signaling pathway has been implicated in several forms of cancer. The pathway, which conveys an anti-apoptotic signal, has been found to be activated in pancreatic adenocarcinoma tissues.

Development

Programmed cell death is an integral part of both plant and animal tissue development. Development of an organ or tissue is often preceded by the extensive division and differentiation of a particular cell, the resultant mass is then "pruned" into the correct form by apoptosis. Unlike cellular death caused by injury, apoptosis results in cell shrinkage and fragmentation. This allows the cells to be efficiently phagocytosed and their components reused without releasing potentially harmful intracellular substances (such as hydrolytic enzymes, for example) into the surrounding tissue.

Research on chick embryos has suggested how selective cell proliferation, combined with selective apoptosis, sculpts developing tissues in vertebrates. During vertebrate embryo development, structures called the notochord and the floor plate secrete a gradient of the signaling molecule (Shh), and it is this gradient that directs cells to form patterns in the embryonic neural tube: cells that receive Shh in a receptor in their membranes called Patched1 (Ptc1) survive and proliferate; but, in the absence of Shh, one of the ends of this same Ptc1 receptor (the carboxyl-terminal, inside the membrane) is cleaved by caspase-3, an action that exposes an apoptosis-producing domain.

During development, apoptosis is tightly regulated and different tissues use different signals for inducing apoptosis. In birds, bone morphogenetic proteins (BMP) signaling is used to induce apoptosis in the interdigital tissue. In *Drosophila* flies, steroid hormones regulate cell death. Developmental cues can also induce apoptosis, such as the sex-specific cell death of hermaphrodite specific neurons in *C. elegans* males through low *TRA-1* transcription factor activity (*TRA-1* helps prevent cell death).

Lymphocyte Interactions

The development of B lymphocytes and the development of T lymphocytes in the human body is a complex process that effectively creates a large pool of diverse cells to begin with, then weeds out those potentially damaging to the body. Apoptosis is the mechanism by which the body removes both the ineffective and the potentially-damaging immature cells, and in T-cells is initiated by the withdrawal of survival signals.

Cytotoxic T-cells are able to directly induce apoptosis in cells by opening up pores in the target's membrane and releasing chemicals that bypass the normal apoptotic pathway. The pores are created by the action of secreted perforin, and the granules contain granzyme B, a serine protease that activates a variety of caspases by cleaving aspartate residues.

Process

The process of apoptosis is controlled by a diverse range of cell signals, which may originate either extracellularly (*extrinsic inducers*) or intracellularly (*intrinsic inducers*). Extracellular signals may include hormones, growth factors, nitric oxide or cytokines, and therefore must either cross the plasma membrane or transduce to effect a response. These signals may positively or negatively induce apoptosis; in this context the binding and subsequent initiation of apoptosis by a molecule is termed positive, whereas the active repression of apoptosis by a molecule is termed negative.

Intracellular apoptotic signalling is a response initiated by a cell in response to stress, and may ultimately result in cell suicide. The binding of nuclear receptors by glucocorticoids, heat, radiation, nutrient deprivation, viral infection, and hypoxia are all factors that can lead to the release of intracellular apoptotic signals by a damaged cell. A number of cellular components, such as poly ADP ribose polymerase, may also help regulate apoptosis.

Before the actual process of cell death is carried out by enzymes, apoptotic signals must be connected to the actual death pathway by way of regulatory proteins. This step allows apoptotic signals to either culminate in cell death, or be aborted should the cell no longer need to die. Several proteins are involved, however two main methods of achieving regulation have been identified; targeting mitochondria functionality, or directly transducing the signal via *adapter proteins* to the apoptotic mechanisms. The whole preparation process requires energy and functioning cell machinery.

Mitochondrial Regulation

The mitochondria are essential to multicellular life. Without them, a cell ceases to respire aerobically and quickly dies - a fact exploited by some apoptotic pathways. Apoptotic proteins that target mitochondria affect them in different ways; they may cause mitochondrial swelling through the formation of membrane pores, or they may increase the permeability of the mitochondrial membrane and cause apoptotic effectors to leak out. There is also a growing body of evidence that indicates that nitric oxide (NO) is able to induce apoptosis by helping to dissipate the membrane potential of mitochondria and therefore make it more permeable.

Mitochondrial proteins known as SMACs (second mitochondria-derived activator of caspases) are released into the cytosol following an increase in permeability. SMAC binds to *inhibitor of apoptosis proteins* (IAPs) and deactivates them, preventing the IAPs from arresting the apoptotic process and therefore allowing apoptosis to proceed. IAP also normally suppresses the activity of a group of cysteine proteases called caspases, which carry out the degradation of the cell, therefore the actual degradation enzymes can be seen to be indirectly regulated by mitochondrial permeability.

Cytochrome c is also released from mitochondria due to formation of a channel, MAC, in the outer mitochondrial membrane, and serves a regulatory function as it precedes morphological change associated with apoptosis. Once cytochrome c is released it binds with *Apaf-1* and ATP, which then bind to *pro-caspase-9* to create a protein complex known as an apoptosome. The apoptosome cleaves the pro-caspase to its active form of caspase-9, which in turn activates the effector *caspase-3*.

MAC is itself subject to regulation by various proteins, such as those encoded by the mammalian *Bcl-2* family of anti-apoptopic genes, the homologs of the *ced-9* gene found in *C. elegans*. *Bcl-2* proteins are able to promote or inhibit apoptosis either by direct action on MAC or indirectly through other proteins. It is important to note that the actions of some *Bcl-2* proteins are able to halt apoptosis even if cytochrome c has been released by the mitochondria.

Direct Signal Transduction

Two important examples of the direct initiation of apoptotic mechanisms in mammals include the *TNF-induced* (tumour necrosis factor) model and the *Fas-Fas ligand-mediated* model, both involving receptors of the *TNF receptor* (TNFR) family coupled to extrinsic signals.

TNF is a cytokine produced mainly by activated macrophages, and is the major extrinsic mediator of apoptosis. Most cells in the human body have two receptors for TNF: *TNF-R1* and *TNF-R2*. The binding of TNF to *TNF-R1* has been shown to initiate the pathway that leads to caspase activation via the intermediate membrane proteins *TNF receptor-associated death domain* (TRADD) and *Fas-associated death domain protein* (FADD). Binding of this receptor can also indirectly lead to the activation of transcription factors involved in cell survival and inflammatory responses. The link between TNF and apoptosis shows why an abnormal production of TNF plays a fundamental role in several human diseases, especially in autoimmune diseases.

The Fas receptor (also known as *Apo-1* or *CD95*) binds the Fas ligand (FasL), a transmembrane protein part of the TNF family. The interaction between Fas and FasL results in the formation of the *death-inducing signaling complex* (DISC), which contains the FADD, caspase-8 and caspase-10. In some types of cells (type I), processed caspase-8 directly activates other members of the caspase family, and triggers the execution of apoptosis. In other types of cells (type II), the *Fas*-DISC starts a feedback loop that spirals into increasing release of pro-apoptotic factors from mitochondria and the amplified activation of caspase-8.

Following *TNF-R1* and *Fas* activation in mammalian cells a balance between pro-apoptotic (BAX, BID, BAK, or BAD) and anti-apoptotic (*Bcl-Xl* and *Bcl-2*) members of the *Bcl-2* family is established. This balance is the proportion of pro-apoptotic homodimers that form in the outer-membrane of the mitochondrion. The pro-apoptotic homodimers are required to make the mitochondrial membrane permeable for the release of caspase activators such as cytochrome c and SMAC. Control of pro-apoptotic proteins under normal cell conditions of non-apoptotic cells is incompletely understood, but it has been found that a mitochondrial outer-membrane protein, VDAC2, interacts with BAK to keep this potentially-lethal apoptotic effector under control. When the death signal is received, products of the activation cascade displace VDAC2 and BAK is able to be activated.

Execution

Although many pathways and signals lead to apoptosis, there is only one mechanism that actually causes the death of the cell in this process; after the appropriate stimulus has been received by the cell and the necessary controls exerted, a cell will undergo the organized degradation of cellular organelles by activated proteolytic caspases. A cell undergoing apoptosis shows a characteristic morphology that can be observed with a microscope:

— Cell shrinkage and rounding due to the breakdown of the proteinaceous cytoskeleton by caspases.

— The cytoplasm appears dense, and the organelles appear tightly packed.

— Chromatin undergoes condensation into compact patches against the nuclear envelope in a process known as pyknosis, a hallmark of apoptosis.

The nuclear envelope becomes discontinuous and the DNA inside it is fragmented in a process referred to as karyorrhexis. The nucleus breaks into several discrete *chromatin bodies* or *nucleosomal units* due to the degradation of DNA.

— The cell membrane shows irregular buds known as blebs.

— The cell breaks apart into several vesicles called *apoptotic bodies*, which are then phagocytosed.

Apoptosis progresses quickly and its products are quickly removed, making it difficult to detect or visualize. During karyorrhexis, endonuclease activation leaves short DNA fragments, regularly spaced in size. These give a characteristic "laddered" appearance on agar gel after electrophoresis. Tests for DNA laddering differentiate apoptosis from ischemic or toxic cell death.

Removal of Dead Cells

Dying cells that undergo the final stages of apoptosis display phagocytotic molecules, such as phosphatidylserine, on their cell surface. Phosphatidylserine is normally found on the cytosolic surface of the plasma membrane, but is redistributed during apoptosis to the extracellular surface by a hypothetical protein known as scramblase. These molecules mark the cell for phagocytosis by cells possessing the appropriate receptors, such as macrophages. Upon recognition, the phagocyte reorganizes its cytoskeleton for engulfment of the cell. The removal of dying cells by phagocytes occurs in an orderly manner without eliciting an inflammatory response.

IMPLICATION IN DISEASE

Defective Apoptotic Pathways

The many different types of apoptotic pathways contain a multitude of different biochemical components, many of them not yet understood. As a pathway is more or less sequential in nature it is a victim of causality; removing or modifying one component leads to an effect in another. In a living organism this can have disastrous effects, often in the form of disease or disorder. A discussion of every disease caused by modification of the various apoptotic pathways would be impractical, but the concept overlying each one is the same: the normal functioning of the pathway has been disrupted in such a way as to impair the ability of the cell to undergo normal apoptosis. This results in a cell that lives past its "use-by-date" and is able to replicate and pass on any faulty machinery to its progeny, increasing the likelihood of the cell becoming cancerous or diseased.

A recently-described example of this concept in action can be seen in the development of a lung cancer called NCI-H460. The *X-linked inhibitor of apoptosis protein* (XIAP) is overexpressed in cells of the H460 cell line. XIAPs bind to the processed form of caspase-9, and suppress the activity of apoptotic activator cytochrome c, therefore overexpression leads to a decrease in the amount of pro-apoptotic agonists. As a consequence, the balance of anti-apoptotic and pro-apoptotic effectors is upset in favour of the former, and the damaged cells continue to replicate despite being directed to die.

p53 Disregulation

The tumor-suppressor protein p53 accumulates when DNA is damaged due to a chain of biochemical reactions. Part of this pathway includes interferon-alpha and interferon-beta, which induce transcription of the *p53* gene and result in the increase of p53 protein level and enhancement of cancer cell-apoptosis. p53 prevents the cell from replicating by stopping the cell cycle at G1, or interphase, to give the cell time to repair, however it will induce apoptosis if damage is extensive and repair efforts fail. Any disruption to the regulation of the *p53* or interferon genes will result in impaired apoptosis and the possible formation of tumors.

HIV Progression

The progression of the human immunodeficiency virus (HIV) to AIDS is primarily due to the depletion of CD4+ T-helper lymphocytes, which leads to a compromised immune system. One of the mechanisms by which T-helper cells are depleted is apoptosis, which can be the end-product of multiple biochemical pathways:

- — HIV enzymes inactivate anti-apoptotic *Bcl-2* and simultaneously activate pro-apoptotic *procaspase-8*. This does not directly cause cell death but primes the cell for apoptosis should the appropriate signal be received.
- — HIV products may increase levels of cellular proteins which have a promotive effect on Fas-mediated apoptosis.
- — HIV proteins decrease the amount of CD4 glycoprotein marker present on the cell membrane.

Released viral particles and proteins present in extracellular fluid are able to induce apoptosis in nearby "bystander" T-helper cells.

HIV decreases the production of molecules involved in marking the cell for apoptosis, giving the virus time to replicate and continue releasing apoptotic agents and virions into the surrounding tissue.

The infected CD4+ cell may also receive the death signal from a cytotoxic T cell, leading to apoptosis.

In addition to apoptosis, infected cells may also die as a direct consequence of the viral infection.

Viral Infection

Viruses can trigger apoptosis of infected cells via a range of mechanisms including:

- — Receptor binding.
- — Activation of protein kinase R (PKR).
- — Interaction with p53.
- — Expression of viral proteins coupled to MHC proteins on the surface of the infected cell, allowing recognition by cells of the immune system (such as Natural Killer and cytotoxic T cells) that then induce the infected cell to undergo apoptosis.

Most viruses encode proteins that can inhibit apoptosis. Several viruses encode viral homologs of Bcl-2. These homologs can inhibit pro-apoptotic proteins such as BAX and BAK, which are essential for the activation of apoptosis. Examples of viral Bcl-2 proteins include the Epstein-Barr virus BHRF1 protein and the adenovirus E1B 19K protein. Some viruses express caspase inhibitors that inhibit caspase activity and an example is the CrmA protein of cowpox viruses. Whilst a number of viruses can block the effects of TNF and Fas. For example the M-T2 protein of myxoma viruses can bind TNF preventing it from binding the TNF receptor and inducing a response.

Furthermore, many viruses express p53 inhibitors that can bind p53 and inhibit its transcriptional transactivation activity. Consequently p53 cannot induce apoptosis since it cannot induce the expression of pro-apoptotic proteins. The adenovirus E1B-55K protein and the hepatitis B virus HBx protein are examples of viral proteins that can perform such a function.

Interestingly, viruses can remain intact from apoptosis particularly in the latter stages of infection. They can be exported in the *apoptotic bodies* that pinch off from the surface of the dying cell and the fact that they are engulfed by phagocytes prevents the initiation of a host response. This favours the spread of the virus.

AUTOPHAGY

In cell biology, autophagy, or autophagocytosis, is a catabolic process involving the degradation of a cell's own components through the lysosomal machinery. It is a tightly-regulated process that plays a normal part in cell growth, development, and homeostasis, helping to maintain a balance between the synthesis, degradation, and subsequent recycling of cellular products. It is a major mechanism by which a starving cell reallocates nutrients from unnecessary processes to more-essential processes.

A variety of autophagic processes exist, all having in common the degradation of intracellular components via the lysosome. The most well-known mechanism of autophagy involves the formation of a membrane around a targeted region of the cell, separating the contents from the rest of the cytoplasm. The resultant vesicle then fuses with a lysosome and subsequently degrades the contents.

It was first described in the 1960s, but many questions about the actual processes and mechanisms involved still remain to be elucidated. Its role in disease is not well categorised; it may help to halt the progression of some diseases and play a protective role against infection by intracellular pathogens; however, in some situations, it may actually contribute to the development of a disease

Autophagy can be broadly separated into three types: *macroautophagy*, *microautophagy* and *chaperone-mediated autophagy*. Macroautophagy involves the formation of a de-novo-formed membrane sealing on itself to engulf cytosolic components (proteins and/or whole organelles), which are degraded after its fusion with the the lysosome, whereas microautophagy is the direct invagination of materials into the lysosome. Specific types of autophagy include:

— Chaperone-mediated autophagy, a term used to describe the degradation of specific cytosolic proteins marked with a specific peptide sequence. Chaperone molecules bind to and transport marked proteins to the lysosome via a receptor complex.

— Pexophagy, autophagy selective for degradation of peroxisomes, which can be separated into *macropexophagy* and *micropexophagy*.

— Mitophagy, autophagy selective for degradation of mitochondria is called, which can be separated in *macromitophagy* and *micromitophagy*.

— Xenophagy, autophagy selective for degradation of intracellular bacteria and viruses.

Process

Macroautophagy is the sequestration of organelles and long-lived proteins in a double-membrane vesicle, called an *autophagosome* or *autophagic vacuole (AV)*, inside the cell. Autophagosomes form from the elongation of small membrane structures known as *autophagosome precursors*. The formation of autophagosomes is initiated by class III phosphoinositide 3-kinase and autophagy-related gene (Atg) 6 (also known as Beclin-1). In addition, two further systems are involved, composed of the ubiquitin-like protein Atg8 (also known as light chain (LC)3) and the Atg4 protease on the one hand and the Atg12-Atg5-Atg16 complex on the other. The outer membrane of the autophagosome fuses in the cytoplasm with a lysosome to form an *autolysosome* or *autophagolysosome* where their contents are degraded via acidic lysosomal hydrolases.

Microautophagy, on the other hand, happens when lysosomes directly engulf cytoplasm by invaginating, protrusion, and/or septation of the lysosomal limiting membrane. In Chaperone-mediated autophagy, or CMA, only those proteins that have a consensus peptide sequence get recognized by the binding of a hsc70-containing chaperone/co-chaperone complex. This CMA substrate/chaperone complex then moves to the lysosomes, where the CMA receptor lysosome-associated membrane protein type-2A (LAMP-2A) recognizes it; the protein is unfolded and translocated across the lysosome membrane assisted by the lysosomal hsc70 on the other side. CMA differs from macroautophagy and microautophagy in two main ways:

— The substrates are translocated across the lysosome membrane on a one-by-one basis, whereas in the macroautophagy and microautophagy the substrates are engulfed or sequestered in-bulk.

— CMA is very selective in what it degrades and can degrade only certain proteins and not organelles.

— Autophagy is part of everyday normal cell growth and development wherein mTOR plays an important regulatory role.

Functions

Nutrient starvation

During nutrient starvation, increased levels of autophagy lead to the breakdown of non-vital components and the release of nutrients, ensuring that vital processes can continue. Mutant yeast cells that have a reduced autophagic capability rapidly perish in nutrition-deficient conditions. A gene known as *Atg7* has been implicated in nutrient-mediated autophagy, as mice studies have shown that starvation-induced autophagy was impaired in *Atg7*-deficient mice.

Infection

Autophagy plays a role in the destruction of some bacteria within the cell. Intracellular pathogens such as *Mycobacterium tuberculosis* persist within cells and block the normal actions taken by the cell to rid itself of it. Stimulating autophagy in infected cells overcomes the block and helps to rid the cell of pathogens.

REFERENCES

Jun Ito and Hiroo Fukuda, "ZEN1 Is a Key Enzyme in the Degradation of Nuclear DNA during Programmed Cell Death of Tracheary Elements", *The Plant Cell*, December 2002.

Mazal Solomon, et al., "The Involvement of Cysteine Proteases and Protease Inhibitor Genes in the Regulation of Programmed Cell Death in Plants", *The Plant Cell*, Vol. March 1999.

Richard A. Lockshin, Carroll M. Williams, "Programmed cell death—II. Endocrine potentiation of the breakdown of the intersegmental muscles of silkmoths", *Journal of Insect Physiology*, 1964.

9

Cellular Respiration

Cellular respiration describes the metabolic reactions and processes that take place in a cell or across the cell membrane to get biochemical energy from fuel molecules and the release of the cells' waste products. Energy can be released by the oxidation of multiple fuel molecules and is stored as "high-energy" carriers. The reactions involved in respiration are catabolic reactions in metabolism.

Fuel molecules commonly used by cells in respiration include glucose, amino acids and fatty acids, and a common oxidizing agent (electron acceptor) is molecular oxygen (O_2). There are organisms, however, that can respire using other organic molecules as electron acceptors instead of oxygen. Organisms that use oxygen as a final electron acceptor in respiration are described as aerobic, while those that do not are referred to as anaerobic.

The energy released in respiration is used to synthesize molecules that act as a chemical storage of this energy. One of the most widely used compounds in a cell is adenosine triphosphate (ATP) and its stored chemical energy can be used for many processes requiring energy, including biosynthesis, locomotion or transportation of molecules across cell membranes. Because of its ubiquitous nature, ATP is also known as the "universal energy currency", since the amount of it in a cell indicates how much energy is available for energy-consuming processes.

AEROBIC RESPIRATION

Aerobic respiration requires oxygen in order to generate energy (ATP). It is the preferred method of pyruvate breakdown from glycolysis and requires that pyruvate enter the mitochondrion to be fully oxidized by the Krebs cycle. The product of this process is energy in the form of ATP (Adenosine Triphosphate), by substrate-level phosphorylation, NADH and $FADH_2$.

Simplified Reaction:

$$C_6H_{12}O_{6\,(aq)} + 6O_{2\,(g)} \rightarrow 6CO_{2\,(g)} + 6H_2O_{(l)} \rightarrow \Delta H_c\ \text{-2880 kJ}$$

The reducing potential of NADH and $FADH_2$ is converted to more ATP through an electron transport chain with oxygen as the "terminal electron acceptor". Most of the ATP produced by aerobic cellular respiration is made by oxidative phosphorylation. This works by the energy released in the consumption of pyruvate being used to create a chemiosmotic potential by pumping protons across a membrane. This potential is then used to drive ATP synthase and produce ATP from ADP.

Generally, 38 ATP molecules are formed from aerobic respiration. However, this maximum yield is never quite reached due to losses (leaky membranes) as well as the cost of moving pyruvate and ADP into the mitochondrial matrix. Aerobic metabolism is 19 times more efficient than anaerobic metabolism (which yields 2 mol ATP per 1 mol glucose). They share the initial pathway of glycolysis but aerobic metabolism continues with the Krebs cycle and oxidative phosphorylation. The post glycolytic reactions take place in the mitochondria in eukaryotic cells, and in the cytoplasm in prokaryotic cells.

Glycolysis

Glycolysis is the sequence of reactions that converts glucose into pyruvate with the concomitant production of a relatively small amount of adenosine triphosphate (ATP). It is the initial process of most carbohydrate catabolism, and it serves three principal functions:

The generation of high-energy molecules (ATP and NADH) as cellular energy sources as part of aerobic respiration and anaerobic respiration; that is, in the former process, oxygen is present, and, in the latter, oxygen is not present

Production of pyruvate for the citric acid cycle as part of aerobic respiration The production of a variety of six- and three-carbon intermediate compounds, which may be removed at various steps in the process for other cellular purposes.

As the foundation of both aerobic and anaerobic respiration, glycolysis is the archetype of universal metabolic processes known and occurring (with variations) in many types of cells in nearly all organisms. Glycolysis, through anaerobic respiration, is the main energy source in many prokaryotes, eukaryotic cells devoid of mitochondria (e.g., mature erythrocytes) and eukaryotic cells under low-oxygen conditions (e.g., heavily-exercising muscle or fermenting yeast).

In eukaryotes and prokaryotes, glycolysis takes place within the cytosol of the cell. In plant cells, some of the glycolytic reactions are also found in the Calvin-Benson cycle, which functions inside the chloroplasts. The wide conservation includes the most phylogenetically deep-rooted extant organisms, and thus it is considered to be one of the most ancient metabolic pathways.

The products all have vital cellular uses:

— ATP provides an energy source for many cellular functions.

— NADH + H+ provides reducing power for other metabolic pathways or further ATP synthesis.

— Pyruvate is used in the citric acid cycle in aerobic respiration to produce more ATP, or is converted to other small carbon molecules in anaerobic respiration.

For simple anaerobic fermentations, the metabolism of one molecule of glucose to two molecules of pyruvate has a net yield of two molecules of ATP. Most cells will then carry out further reactions to 'repay' the used NAD^+ and produce a final product of ethanol or lactic acid. Many bacteria use inorganic compounds as hydrogen acceptors to regenerate the NAD^+.

Cells performing aerobic respiration synthesize much more ATP, but not as part of glycolysis. These further aerobic reactions use pyruvate and $NADH^+$ H^+ from glycolysis. Eukaryotic aerobic respiration produces approximately 34 additional molecules of ATP for each glucose molecule, however most of these are produced by a vastly different mechanism to the substrate-level phosphorylation in glycolysis.

The lower energy production, per glucose, of anaerobic respiration relative to aerobic respiration, results in greater flux through the pathway under hypoxic (low-oxygen) conditions, unless alternative sources of anaerobically-oxidizable substrates, such as fatty acids, are found.

The first formal studies of the glycolytic process were initiated in 1860 when Louis Pasteur discovered that microorganisms are responsible for fermentation, and in 1897 when Eduard Buchner found certain cell extracts can cause fermentation. The next major contribution was from Arthur Harden and William Young in 1905 who determined that a heat-sensitive high-molecular-weight subcellular fraction (the enzymes) and a heat-insensitive low-molecular-weight cytoplasm fraction (ADP, ATP and NAD^+ and other cofactors) are required together for fermentation to proceed. The details of the pathway itself were eventually determined by 1940, with a major input from Otto Meyerhof and some years later by Luis Leloir.

The biggest difficulties in determining the intricacies of the pathway were due to the very short lifetime and low steady-state concentrations of the intermediates of the fast glycolytic reactions.

Regulation

Gluconeogenesis

The flux through the glycolytic pathway is adjusted in response to conditions both inside and outside the cell. The rate in liver is regulated to meet major cellular needs: (1) the production of ATP, (2) the provision of building blocks for biosynthetic reactions, and (3) to lower blood glucose, one of the major functions of the liver. When blood sugar falls, glycolysis is halted in liver to allow the reverse process, gluconeogenesis. In glycolysis, the reactions catalyzed by hexokinase, phosphofructokinase, and pyruvate kinase are effectively irreversible in most organisms. In metabolic pathways, such enzymes are potential sites of control, and all three enzymes serve this purpose in glycolysis.

There are several different ways to regulate the activity of an enzyme. An immediate form of control is feedback via allosteric effectors or by covalent modification. A slower form of control is transcriptional regulation that controls the amounts of these important enzymes.

Hexokinase

Hexokinase is inhibited by glucose-6-phosphate (G6P), the product it forms through the ATP-driven phosphorylation. This is necessary to prevent an accumulation of G6P in the cell when flux through the glycolytic pathway is low. Glucose will enter the cell, but, since the hexokinase has reduced activity, it can diffuse back into the blood through the glucose transporter in the plasma membrane.

In animals, regulation of blood glucose levels by the liver is a vital part of homeostasis. In liver cells, extra G6P may be converted to G1P for conversion to glycogen, or it is alternatively converted by glycolysis to acetyl-CoA and then citrate. Excess citrate is exported to the cytosol, where ATP citrate lyase will regenerate acetyl-CoA and OAA.

The acetyl-CoA is then used for fatty acid and cholesterol synthesis, two important ways of utilizing excess glucose when its concentration is high in blood. Liver contains both hexokinase and glucokinase; the latter catalyses the phosphorylation of glucose to G6P and is not inhibited by G6P. Thus it allows glucose to be converted into glycogen, fatty acids, and cholesterol even when hexokinase activity is low. This is important when blood glucose levels are high. During hypoglycemia, the glycogen can be converted back to G6P and then converted to glucose by a liver-specific enzyme glucose 6-phosphatase. This reverse reaction is an important role of liver cells to maintain blood sugars levels during fasting. This is critical for neuron function, since they can use only glucose as an energy source.

Phosphofructokinase

Phosphofructokinase is an important control point in the glycolytic pathway, since it is one of the irreversible steps and has key allosteric effectors, AMP and fructose 2,6-bisphosphate (F2,6BP).

Fructose 2,6-bisphosphate (F2,6BP) is a very potent activator of phosphofructokinase (PFK-1) that is synthesised when F6P is phosphorylated by a second phosphofructokinase (PFK_2). In liver, when blood sugar is low and glucagon elevates cAMP, PFK_2 is phosphorylated by protein kinase A. The phosphorylation inactivates PFK2, and another domain on this protein becomes active as fructose 2,6-bisphosphatase, which converts F2,6BP back to

F6P. Both glucagon and epinephrine cause high levels of cAMP in the liver.

The result of lower levels of liver fructose-2,6-bisphosphate is a decrease in activity of phosphofructokinase and an increase in activity of fructose 1,6-bisphosphatase, so that gluconeogenesis (essentially "glycolysis in reverse") is favored. This is consistent with the role of the liver in such situations, since the response of the liver to these hormones is to release glucose to the blood.

ATP competes with AMP for the allosteric effector site on the PFK enzyme. ATP concentrations in cells are much higher than AMP, typically 100-fold higher, but the concentration of ATP does not change more than about 10% under physiological conditions, whereas a 10% drop in ATP results in a 6-fold increase in AMP. Thus, the relevance of ATP as an allosteric effector is questionable. An increase in AMP is a consequence of a decrease in energy charge in the cell.

Citrate inhibits phosphofructokinase when tested *in vitro* by enhancing the inhibitory effect of ATP. However, it is doubtful that this is a meaningful effect *in vivo*, because citrate in the cytosol is mainly utilized for conversion to acetyl-CoA for fatty acid and cholesterol synthesis.

Pyruvate kinase and phosphoglycerate kinase

Pyruvate kinase and phosphoglycerate kinase catalyze the two substrate-level phosphorylation steps, and produce ATP from ADP. While both of these reactions are exergonic, phosphoglycerate kinase is less exergonic (-18.8 kJ/mol) than pyruvate kinase. Phosphoglycerate kinase helps to "pull along" the endergonic glyceraldehyde phosphate dehydrogenase, and in fact, these enzymes are reversible and also function in gluconeogenesis. In contrast, the strongly exergonic pyruvate kinase is irreversible and thus a prime candidate for regulation.

Post-glycolysis processes

The ultimate fate of pyruvate and NADH produced in glycolysis depends upon the organism and the conditions, most notably the presence or absence of oxygen and other external electron acceptors. In addition, not all carbon entering the pathway leaves as pyruvate and may be extracted at earlier stages to provide carbon compounds for other pathways.

Oxidative decarboxylation of pyruvate

The pyruvate is oxidized to acetyl-CoA and CO_2 by the Pyruvate dehydrogenase complex, a cluster of enzymes—multiple copies of each of three enzymes—located in the mitochondria of eukaryotic cells and in the cytosol of prokaryotes. In the process one molecule of NADH is formed per pyruvate oxidized.

CITRIC ACID CYCLE

The citric acid cycle, also known as the tricarboxylic acid cycle (TCA cycle) or the Krebs cycle, (or rarely, the Szent-Györgyi-Krebs cycle) is a series of enzyme-catalysed chemical reactions of central importance in all living cells that use oxygen as part of cellular respiration. In eukaryotes, the citric acid cycle occurs in the matrix of the mitochondrion. The components and reactions of the citric acid cycle were established by seminal work from both Albert Szent-Györgyi and Hans Krebs.

In aerobic organisms, the citric acid cycle is part of a metabolic pathway involved in the chemical conversion of carbohydrates, fats and proteins into carbon dioxide and water to generate a form of usable energy. Other relevant reactions in the pathway include those in glycolysis and pyruvate oxidation before the citric acid cycle, and oxidative phosphorylation after it. In addition, it provides precursors for many compounds including some amino acids and is therefore functional even in cells performing fermentation.

Two carbons are oxidized to CO_2, and the energy from these reactions is transferred to other metabolic processes by GTP (or ATP), and as electrons in NADH and QH_2. The NADH generated in the TCA cycle may later donate its electrons in oxidative phosphorylation to drive ATP synthesis; $FADH_2$ is covalently attached to succinate dehydrogenase, an enzyme functioning both in the TCA cycle and the mitochondrial electron transport chain in oxidative phosphorylation. $FADH_2$ thereby facilitates transfer of electrons to coenzyme Q, which is the final electron acceptor of the reaction catalyzed by the Succinate:ubiquinone oxidoreductase complex, also acting as an intermediate in the electron transport chain.

Products of the first turn of the cycle are: *one GTP (or ATP), three NADH, one QH_2, two CO_2*. Because two acetyl-CoA molecules are produced from each glucose molecule, two cycles are required per glucose molecule. Therefore, at the end of all cycles, the products are: two GTP, six NADH, two QH_2, and four CO_2

Estimates for the total number of ATP obtained after complete oxidation of one glucose in glycolysis, citric acid cycle, and oxidative phosphorylation given in the literature range from 30-38 molecules of ATP. A recent assessment of the total ATP yield obtained in these distinct reaction cycles, taking into account updated proton-to-ATP ratios, has arrived at an estimate of 29.85 ATP per glucose molecule.

Regulation

The regulation of the TCA cycle is largely determined by substrate availability and product inhibition. NADH, a product of all dehydrogenases in the TCA cycle with the exception of succinate dehydrogenase, inhibits pyruvate dehydrogenase, isocitrate dehydrogenase and a-ketoglutarate dehydrogenase, and also citrate synthase. Acetyl-CoA inhibits pyruvate dehydrogenase, while succinyl-CoA inhibits succinyl-CoA synthase and citrate synthase. When tested in vitro with TCA enzymes, ATP inhibits citrate synthase and a-ketoglutarate dehydrogenase; however, ATP levels do not change more than 10% in vivo between rest and vigorous exercise. There is no known allosteric mechanism that can account for large changes in reaction rate from an allosteric effector whose concentration changes less than 10%.

Calcium is used as a regulator. It activates pyruvate dehydrogenase, isocitrate dehydrogenase and a-ketoglutarate dehydrogenase. This increases the reaction rate of many of the steps in the cycle, and therefore increases flux throughout the pathway.

Citrate is used for feedback inhibition, as it inhibits phosphofructokinase, an enzyme involved in glycolysis that catalyses formation of fructose 1,6-bisphosphate, a precursor of pyruvate. This prevents a constant high rate of flux when there is an accumulation of citrate and a decrease in substrate for the enzyme.

Recent work has demonstrated an important link between intermediates of the citric acid cycle and the regulation of hypoxia inducible factors (HIF). HIF plays a role in the regulation of oxygen haemostasis, and is a transcription factor which targets angiogenesis, vascular remodelling, glucose ulitisation, iron transport and apoptosis. HIF is synthesized consititutively and hydroxylation of at least one of two critical proline residues mediates their interation with the von Hippel Lindau E3 ubiquitin ligase complex which targets them for rapid degradation. This reaction is calalysed by prolyl 4-hydroxylases. Fumarate and succinate have been identified as potent inhibitors of prolyl hydroxylases thus leading to the stabilisation of HIF.

Metabolic Pathways on the TCA cycle

Several catabolic pathways converge on the TCA cycle. Reactions that form intermediates of the TCA cycle in order to replenish them (especially during the scarcity of the intermediates) are called anaplerotic reactions. The citric acid cycle is the third step in carbohydrate catabolism (the breakdown of sugars). Glycolysis breaks glucose (a six-carbon-molecule) down into pyruvate (a three-carbon molecule). In eukaryotes, pyruvate moves into the mitochondria. It is converted into acetyl-CoA by decarboxylation and enters the citric acid cycle.

In protein catabolism, proteins are broken down by protease enzymes into their constituent amino acids. The carbon backbone of these amino acids can become a source of energy by being converted to Acetyl-CoA and entering into the citric acid cycle.

In fat catabolism, triglycerides are hydrolyzed to break them into fatty acids and glycerol. In the liver the glycerol can be converted into glucose via dihydroxyacetone phosphate and glyceraldehyde-3-phosphate by way of gluconeogenesis. In many tissues, especially heart tissue, fatty acids are broken down through a process known as beta oxidation which results in acetyl-CoA which can be used in the citric acid cycle. Sometimes beta oxidation can yield propionyl CoA which can result in further glucose production by gluconeogenesis in the liver.

The citric acid cycle is always followed by oxidative phosphorylation. This process extracts the energy (as electrons) from NADH and QH_2, oxidizing them to NAD^+ and Q, respectively, so that the cycle can continue. Whereas the citric acid cycle does not use oxygen, oxidative phosphorylation does.

The total energy gained from the complete breakdown of one molecule of glucose by glycolysis, the citric acid cycle and oxidative phosphorylation equals about 30 ATP molecules, in eukaryotes. The citric acid cycle is called an amphibolic pathway because it participates in both catabolism and anabolism.

Oxidative Phosphorylation

In eukaryotes, oxidative phosphorylation occurs in the mitochondrial cristae. It comprises the electron transport chain that establishes a proton gradient (chemiosmotic potential) across the inner membrane by oxidizing the NADH produced from the Krebs cycle. ATP is synthesised by the ATP synthase enzyme when the chemiosmotic gradient is used to drive the phosphorylation of ADP.

THEORETICAL YIELDS

The yields in the table below are for one glucose molecule being fully oxidized into carbon dioxide. It is assumed that all the reduced coenzymes are oxidized by the electron transport chain and used for oxidative phosphorylation.

Although there is a theoretical yield of 36-38 ATP molecules per glucose during cellular respiration, such conditions are generally not realized due to losses such as the cost of moving pyruvate (from glycolysis), phosphate, and ADP (substrates for ATP synthesis) into the mitochondria. All are actively transported using carriers that utilise the stored energy in the proton electrochemical gradient.

- The pyruvate carrier is a symporter and the driving force for moving pyruvate into the mitochondria is the movement of protons from the intermembrane space to the matrix.
- The phosphate carrier is an antiporter and the driving force for moving phosphate ions into the mitochondria is the movement of hydroxyls ions from the matrix to the intermembrane space.

— The adenine nucleotide carrier is an antiporter and exchanges ADP and ATP across the inner membrane. The driving force is due to the ATP (-4) having a more negative charge than the ADP (-3) and thus it dissipates some of the electrical component of the proton electrochemical gradient.

The outcome of these transport processes using the proton electrochemical gradient is that more than 3 H^+ are needed to make 1 ATP. Obviously this reduces the theoretical efficiency of the whole process and the likely maximum is closer to 28-30 ATP molecules. In practice the efficiency may be even lower due to the inner membrane of the mitochondria being slightly leaky to protons. Other factors may also dissipate the proton gradient creating an apparently leaky mitochondria. An uncoupling protein known as thermogenin is expressed in some cell types and is a channel that can transport protons. When this protein is active in the inner membrane it short circuits the coupling between the electron transport chain and ATP synthesis. The potential energy from the proton gradient is not used to make ATP but generates heat. This is particularly important in a baby's brown fat, for thermogenesis, and hibernating animals.

ANAEROBIC RESPIRATION

Without oxygen, pyruvate is not metabolized by cellular respiration but undergoes a process of fermentation. The pyruvate is not transported into the mitochondrion, but remains in the cytoplasm, where it is converted to waste products that may be removed from the cell. This serves the purpose of oxidizing the hydrogen carriers so that they can perform glycolysis again and removing the excess pyruvate. This waste product varies depending on the organism. In skeletal muscles, the waste product is lactic acid. This type of fermentation is called lactic acid fermentation. In yeast, the waste products are ethanol and carbon dioxide. This type of fermentation is known as alcoholic or ethanol fermentation. The ATP generated in this process is made by *substrate phosphorylation*, which is phosphorylation that does not involve oxygen.

Anaerobic respiration is less efficient at using the energy from glucose since 2 ATP are produced during anaerobic respiration per glucose, compared to the 30 ATP per glucose produced by aerobic respiration. This is because the waste products of anaerobic respiration still contain plenty of energy. Ethanol, for example, can be used in gasoline (petrol) solutions. Glycolytic ATP, however, is created more quickly. For prokaryotes to continue a rapid growth rate when they are shifted from an aerobic environment to an anaerobic environment, they must increase the rate of the glycolytic reactions. Thus, during short bursts of strenuous activity, muscle cells use anaerobic respiration to supplement the ATP production from the slower aerobic respiration, so anaerobic respiration may be used by a cell even before the oxygen levels are depleted, as is the case in sports that do not require athletes to pace themselves, such as sprinting.

REFERENCES

Rich P.R., "The molecular machinery of Keilin's respiratory chain", *Biochem. Soc. Trans.*, 2003.

Porter R.K, Brand M.D., "Mitochondrial proton conductance and H+/O ratio are independent of electron transport rate in isolated hepatocytes", *Biochemistry Journal*, 1995.

10

Cell Metabolism

Metabolism is the set of chemical reactions that occur in living organisms in order to maintain life. These processes allow organisms to grow and reproduce, maintain their structures, and respond to their environments. Metabolism is usually divided into two categories. Catabolism breaks down large molecules, for example to harvest energy in cellular respiration. Anabolism, on the other hand, uses energy to construct components of cells such as proteins and nucleic acids.

The chemical reactions of metabolism are organized into metabolic pathways, in which one chemical is transformed into another by a sequence of enzymes. Enzymes are crucial to metabolism because they allow organisms to drive desirable but thermodynamically unfavorable reactions by coupling them to favorable ones. Enzymes also allow the regulation of metabolic pathways in response to changes in the cell's environment or signals from other cells.

The metabolism of an organism determines which substances it will find nutritious and which it will find poisonous. For example, some prokaryotes use hydrogen sulfide as a nutrient, yet this gas is poisonous to animals. The speed of metabolism, the metabolic rate, also influences how much food an organism will require.

A striking feature of metabolism is the similarity of the basic metabolic pathways between even vastly different species. For example, the set of carboxylic acids that are best known as the intermediates in the citric acid cycle are present in all organisms, being found in species as diverse as the unicellular bacteria *Escherichia coli* and huge multicellular organisms like elephants. These striking similarities in metabolism are most likely the result of the high efficiency of these pathways, and of their early appearance in evolutionary history.

IMPORTANT BIOCHEMICALS

Most of the structures that make up animals, plants and microbes are made from three basic classes of molecule: amino acids, carbohydrates and lipids (often called fats). As these molecules are vital for life, metabolism focuses on making these molecules, in the construction of cells and tissues, or breaking them down and using them as a source of energy, in the digestion and use of food. Many important biochemicals can be joined together to make polymers such as DNA and proteins. These macromolecules are essential parts of all living organisms.

Amino Acids and Proteins

Proteins are made of amino acids arranged in a linear chain and joined together by peptide bonds. Many proteins are the enzymes that catalyze the chemical reactions in metabolism. Other proteins have structural or mechanical functions, such as the proteins that form the cytoskeleton, a system of scaffolding that maintains the cell shape. Proteins are also important in cell signaling, immune responses, cell adhesion, active transport across membranes and the cell cycle.

Lipids

Lipids are the most diverse group of biochemicals. Their

main structural uses are as part of biological membranes such as the cell membrane, or as a source of energy. Lipids are usually defined as hydrophobic or amphipathic biological molecules that will dissolve in organic solvents such as benzene or chloroform. The fats are a large group of compounds that contain fatty acids and glycerol; a glycerol molecule attached to three fatty acid esters is a triacylglyceride. Several variations on this basic structure exist, including alternate backbones such as sphingosine in the sphingolipids, and hydrophilic groups such as phosphate in phospholipids. Steroids such as cholesterol are another major class of lipids that are made in cells.

Carbohydrates

Carbohydrates are straight-chain aldehydes or ketones with many hydroxyl groups that can exist as straight chains or rings. Carbohydrates are the most abundant biological molecules, and fill numerous roles, such as the storage and transport of energy (starch, glycogen) and structural components (cellulose in plants, chitin in animals). The basic carbohydrate units are called monosaccharides and include galactose, fructose, and most importantly glucose. Monosaccharides can be linked together to form polysaccharides in almost limitless ways.

Nucleotides

The polymers DNA and RNA are long chains of nucleotides. These molecules are critical for the storage and use of genetic information, through the processes of transcription and protein biosynthesis. This information is protected by DNA repair mechanisms and propagated through DNA replication. A few viruses have an RNA genome, for example HIV, which uses reverse transcription to create a DNA template from its viral RNA genome. RNA in ribozymes such as spliceosomes and ribosomes is similar to enzymes as it can catalyze chemical reactions. Individual nucleosides are made by attaching a nucleobase to a ribose sugar. These bases are heterocyclic rings containing nitrogen, classified as purines or pyrimidines. Nucleotides also act as coenzymes in metabolic group transfer reactions.

Coenzymes

Metabolism involves a vast array of chemical reactions, but most fall under a few basic types of reactions that involve the transfer of functional groups. This common chemistry allows cells to use a small set of metabolic intermediates to carry chemical groups between different reactions. These group-transfer intermediates are called coenzymes. Each class of group-transfer reaction is carried out by a particular coenzyme, which is the substrate for a set of enzymes that produce it, and a set of enzymes that consume it. These coenzymes are therefore continuously being made, consumed and then recycled.

One central coenzyme is adenosine triphosphate (ATP), the universal energy currency of cells. This nucleotide is used to transfer chemical energy between different chemical reactions. There is only a small amount of ATP in cells, but as it is continuously regenerated, the human body can use about its own weight in ATP per day. ATP acts as a bridge between catabolism and anabolism, with catabolic reactions generating ATP and anabolic reactions consuming it. It also serves as a carrier of phosphate groups in phosphorylation reactions.

A vitamin is an organic compound needed in small quantities that cannot be made in the cells. In human nutrition, most vitamins function as coenzymes after modification; for example, all water-soluble vitamins are phosphorylated or are coupled to nucleotides when they are used in cells. Nicotinamide adenine dinucleotide (NADH), a derivative of vitamin B3 (niacin), is an important coenzyme that acts as a hydrogen acceptor. Hundreds of separate types of dehydrogenases remove electrons from their substrates and reduce NAD+ into NADH. This reduced form of the coenzyme is then a substrate for any of the reductases in the cell that need to reduce their substrates.

Nicotinamide adenine dinucleotide exists in two related forms in the cell, NADH and NADPH. The NAD+/NADH form is more important in catabolic reactions, while NADP+/NADPH is used in anabolic reactions.

Minerals and Cofactors

Inorganic elements play critical roles in metabolism; some are abundant (e.g. sodium and potassium) while others function at minute concentrations. About 99% of mammals' mass are the elements carbon, nitrogen, calcium, sodium, chlorine, potassium, hydrogen, phosphorus, oxygen and sulfur. The organic compounds (proteins, lipids and carbohydrates) contain the majority of the carbon and nitrogen and most of the oxygen and hydrogen is present as water.

The abundant inorganic elements act as ionic electrolytes. The most important ions are sodium, potassium, calcium, magnesium, chloride, phosphate, and the organic ion bicarbonate. The maintenance of precise gradients across cell membranes maintains osmotic pressure and pH. Ions are also critical for nerves and muscles, as action potentials in these tissues are produced by the exchange of electrolytes between the extracellular fluid and the cytosol. Electrolytes enter and leave cells through proteins in the cell membrane called ion channels. For example, muscle contraction depends upon the movement of calcium, sodium and potassium through ion channels in the cell membrane and T-tubules.

The transition metals are usually present as trace elements in organisms, with zinc and iron being most abundant. These metals are used in some proteins as cofactors and are essential for the activity of enzymes such as catalase and oxygen-carrier proteins such as hemoglobin. These cofactors are bound tightly to a specific protein; although enzyme cofactors can be modified during catalysis, cofactors always return to their original state after catalysis has taken place. The metal micronutrients are taken up into organisms by specific transporters and bound to storage proteins such as ferritin or metallothionein when not being used.

CATABOLISM

Catabolism is the set of metabolic processes that break down large molecules. These include breaking down and oxidising food molecules. The purpose of the catabolic reactions is to provide the energy and components needed by anabolic reactions. The exact nature of these catabolic reactions differ from organism to organism, with organic molecules being used as a source of energy in organotrophs, while lithotrophs use inorganic substrates and phototrophs capture sunlight as chemical energy. However, all these different forms of metabolism depend on redox reactions that involve the transfer of electrons from reduced donor molecules such as organic molecules, water, ammonia, hydrogen sulfide or ferrous ions to acceptor molecules such as oxygen, nitrate or sulfate.

In animals these reactions involve complex organic molecules being broken down to simpler molecules, such as carbon dioxide and water. In photosynthetic organisms such as plants and cyanobacteria, these electron-transfer reactions do not release energy, but are used as a way of storing energy absorbed from sunlight.

The most common set of catabolic reactions in animals can be separated into three main stages. In the first, large organic molecules such as proteins, polysaccharides or lipids are digested into their smaller components outside cells. Next, these smaller molecules are taken up by cells and converted to yet smaller molecules, usually acetyl coenzyme A (CoA), which releases some energy. Finally, the acetyl group on the CoA is oxidised to water and carbon dioxide in the citric acid cycle and electron transport chain, releasing the energy that is stored by reducing the coenzyme nicotinamide adenine dinucleotide (NAD+) into NADH.

Digestion

Macromolecules such as starch, cellulose or proteins cannot be rapidly taken up by cells and need to be broken into their smaller units before they can be used in cell metabolism. Several common classes of enzymes digest these polymers. These digestive enzymes include proteases that digest proteins into amino acids, as well as glycoside hydrolases that digest polysaccharides into monosaccharides.

Microbes simply secrete digestive enzymes into their surroundings, while animals only secrete these enzymes from specialized cells in their guts. The amino acids or sugars released by these extracellular enzymes are then pumped into cells by specific active transport proteins.

Energy from Organic Compounds

Carbohydrate catabolism is the breakdown of carbohydrates into smaller units. Carbohydrates are usually taken into cells once they have been digested into monosaccharides. Once inside, the major route of breakdown is glycolysis, where sugars such as glucose and fructose are converted into pyruvate and some ATP is generated. Pyruvate is an intermediate in several metabolic pathways, but the majority is converted to acetyl-CoA and fed into the citric acid cycle. Although some more ATP is generated in the citric acid cycle, the most important product is NADH, which is made from NAD+ as the acetyl-CoA is oxidized.

This oxidation releases carbon dioxide as a waste product. In anaerobic conditions, glycolysis produces lactate, through the enzyme lactate dehydrogenase re-oxidizing NADH to NAD+ for re-use in glycolysis. An alternative route for glucose breakdown is the pentose phosphate pathway, which reduces the coenzyme NADPH and produces pentose sugars such as ribose, the sugar component of nucleic acids.

Fats are catabolised by hydrolysis to free fatty acids and glycerol. The glycerol enters glycolysis and the fatty acids are broken down by beta oxidation to release acetyl-CoA, which then is fed into the citric acid cycle. Fatty acids release more energy upon oxidation than carbohydrates because carbohydrates contain more oxygen in their structures.

Amino acids are either used to synthesize proteins and other biomolecules, or oxidized to urea and carbon dioxide as a source of energy. The oxidation pathway starts with the removal of the amino group by a transaminase. The amino group is fed into the urea cycle, leaving a deaminated carbon skeleton in the form of a keto acid. Several of these keto acids are intermediates in the citric acid cycle, for example the deamination of glutamate forms a-ketoglutarate. The glucogenic amino acids can also be converted into glucose, through gluconeogenesis.

Energy Transformations

Oxidative phosphorylation

In oxidative phosphorylation, the electrons removed from food molecules in pathways such as the citric acid cycle are transferred to oxygen and the energy released used to make

ATP. This is done in eukaryotes by a series of proteins in the membranes of mitochondria called the electron transport chain. In prokaryotes, these proteins are found in the cell's inner membrane. These proteins use the energy released from passing electrons from reduced molecules like NADH onto oxygen to pump protons across a membrane.

Pumping protons out of the mitochondria creates a proton concentration difference across the membrane and generates an electrochemical gradient. This force drives protons back into the mitochondrion through the base of an enzyme called ATP synthase. The flow of protons makes the stalk subunit rotate, causing the active site of the synthase domain to change shape and phosphorylate adenosine diphosphate - turning it into ATP.

Energy from inorganic compounds

Chemolithotrophy is a type of metabolism found in prokaryotes where energy is obtained from the oxidation of inorganic compounds. These organisms can use hydrogen, reduced sulfur compounds (such as sulfide, hydrogen sulfide and thiosulfate), ferrous iron (FeII) or ammonia as sources of reducing power and they gain energy from the oxidation of these compounds with electron acceptors such as oxygen or nitrite. These microbial processes are important in global biogeochemical cycles such as acetogenesis, nitrification and denitrification and are critical for soil fertility.

Energy from light

The energy in sunlight is captured by plants, cyanobacteria, purple bacteria, green sulfur bacteria and some protists. This process is often coupled to the conversion of carbon dioxide into organic compounds, as part of photosynthesis, which is discussed below. The energy capture and carbon fixation systems can however operate separately in prokaryotes, as purple bacteria and green sulfur bacteria can use sunlight as a source of energy, while switching between carbon fixation and the fermentation of organic compounds.

The capture of solar energy is a process that is similar in principle to oxidative phosphorylation, as it involves energy being stored as a proton concentration gradient and this proton motive force then driving ATP synthesis. The electrons needed to drive this electron transport chain come from light-gathering proteins called photosynthetic reaction centres. These structures are classed into two types depending on the type of photosynthetic pigment present, with most photosynthetic bacteria only having one type of reaction center, while plants and cyanobacteria have two.

In plants, photosystem II uses light energy to remove electrons from water, releasing oxygen as a waste product. The electrons then flow to the cytochrome b6f complex, which uses their energy to pump protons across the thylakoid membrane in the chloroplast. These protons move back through the membrane as they drive the ATP synthase, as before. The electrons then flow through photosystem I and can then either be used to reduce the coenzyme NADP+, for use in the Calvin cycle which is discussed below, or recycled for further ATP generation.

ANABOLISM

Anabolism is the set of constructive metabolic processes where the energy released by catabolism is used to synthesize complex molecules. In general, the complex molecules that make up cellular structures are constructed step-by-step from small and simple precursors. Anabolism involves three basic stages. Firstly, the production of precursors such as amino acids, monosaccharides, isoprenoids and nucleotides, secondly, their activation into reactive forms using energy from ATP, and thirdly, the assembly of these precursors into complex molecules such as proteins, polysaccharides, lipids and nucleic acids.

Organisms differ in how many of the molecules in their cells they can construct for themselves. Autotrophs such as plants can construct the complex organic molecules in cells such as polysaccharides and proteins from simple molecules like carbon dioxide and water. Heterotrophs, on the other hand, require a source of more complex substances, such as monosaccharides and amino acids, to produce these complex molecules. Organisms can be further classified by ultimate source of their energy: photoautotrophs and photoheterotrophs obtain energy from light, whereas chemoautotrophs and chemoheterotrophs obtain energy from inorganic oxidation reactions.

Carbon Fixation

Photosynthesis is the synthesis of carbohydrates from sunlight, carbon dioxide (CO_2) and water, with oxygen produced as a waste product. This process uses the ATP and NADPH produced by the photosynthetic reaction centres, as described above, to convert CO_2 into glycerate 3-phosphate, which can then be converted into glucose. This carbon-fixation reaction is carried out by the enzyme RuBisCO as part of the Calvin – Benson cycle. Three types of photosynthesis occur in plants, C3 carbon fixation, C4 carbon fixation and CAM photosynthesis. These differ by the route that carbon dioxide takes to the Calvin cycle, with C3 plants fixing CO_2 directly, while C4 and CAM photosynthesis incorporate the CO_2 into other compounds first, as adaptations to deal with intense sunlight and dry conditions.

In photosynthetic prokaryotes the mechanisms of carbon fixation are more diverse. Here, carbon dioxide can be fixed by the Calvin – Benson cycle, a reversed citric acid cycle, or the carboxylation of acetyl-CoA. Prokaryotic chemoautotrophs also fix CO_2 through the Calvin – Benson cycle, but use energy from inorganic compounds to drive the reaction.

Carbohydrates and Glycans

In carbohydrate anabolism, simple organic acids can be converted into monosaccharides such as glucose and then used to assemble polysaccharides such as starch. The generation of glucose from compounds like pyruvate, lactate, glycerol, glycerate 3-phosphate and amino acids is called gluconeogenesis. Gluconeogenesis converts pyruvate to glucose-6-phosphate through a series of intermediates, many of which are shared with glycolysis. However, this pathway is not simply glycolysis run in reverse, as several steps are catalyzed by non-glycolytic enzymes. This is important as it allows the formation and breakdown of glucose to be regulated separately and prevents both pathways from running simultaneously in a futile cycle.

Although fat is a common way of storing energy, in vertebrates such as humans the fatty acids in these stores cannot be converted to glucose through gluconeogenesis as these organisms cannot convert acetyl-CoA into pyruvate. As a result, after long-term starvation, vertebrates need to produce ketone bodies from fatty acids to replace glucose in tissues such as the brain that cannot metabolize fatty acids. In other organisms such as plants and bacteria, this metabolic problem is solved using the glyoxylate cycle, which bypasses the decarboxylation step in the citric acid cycle and allows the transformation of acetyl-CoA to oxaloacetate, where it can be used for the production of glucose.

Polysaccharides and glycans are made by the sequential addition of monosaccharides by glycosyltransferase from a reactive sugar-phosphate donor such as uridine diphosphate glucose (UDP-glucose) to an acceptor hydroxyl group on the growing polysaccharide. As any of the hydroxyl groups on the ring of the substrate can be acceptors, the polysaccharides produced can have straight or branched structures. The polysaccharides produced can have structural or metabolic functions themselves, or be transferred to lipids and proteins by enzymes called oligosaccharyltransferases.

Fatty Acids, Isoprenoids and Steroids

Fatty acids are made by fatty acid synthases that polymerize and then reduce acetyl-CoA units. The acyl chains in the fatty acids are extended by a cycle of reactions that add the actyl group, reduce it to an alcohol, dehydrate it to an alkene group and then reduce it again to an alkane group. The enzymes of fatty acid biosynthesis are divided into two groups, in animals and fungi all these fatty acid synthase reactions are carried out by a single multifunctional type I protein, while in plant plastids and bacteria separate type II enzymes perform each step in the pathway.

Terpenes and isoprenoids are a large class of lipids that include the carotenoids and form the largest class of plant natural products. These compounds are made by the assembly and modification of isoprene units donated from the reactive precursors isopentenyl pyrophosphate and dimethylallyl pyrophosphate. These precursors can be made in different ways. In animals and archaea, the mevalonate pathway produces these compounds from acetyl-CoA, while in plants and bacteria the non-mevalonate pathway uses pyruvate and glyceraldehyde 3-phosphate as substrates. One important reaction that uses these activated isoprene donors is steroid biosynthesis. Here, the isoprene units are joined together to make squalene and then folded up and formed into a set of rings to make lanosterol. Lanosterol can then be converted into other steroids such as cholesterol and ergosterol.

Proteins

Organisms vary in their ability to synthesize the 20 common amino acids. Most bacteria and plants can synthesize all twenty, but mammals can synthesize only the ten nonessential amino acids. Thus, the essential amino acids must be obtained from food. All amino acids are synthesized from intermediates in glycolysis, the citric acid cycle, or the pentose phosphate pathway. Nitrogen is provided by glutamate and glutamine. Amino acid synthesis depends on the formation of the appropriate alpha-keto acid, which is then transaminated to form an amino acid.

Amino acids are made into proteins by being joined together in a chain by peptide bonds. Each different protein has a unique sequence of amino acid residues: this is its primary structure. Just as the letters of the alphabet can be combined to form an almost endless variety of words, amino acids can be linked in varying sequences to form a huge variety of proteins. Proteins are made from amino acids that have been activated by attachment to a transfer RNA molecule through an ester bond. This aminoacyl-tRNA precursor is produced in an ATP-dependent reaction carried out by an aminoacyl tRNA synthetase. This aminoacyl-tRNA is then a substrate for the ribosome, which joins the amino acid onto the elongating protein chain, using the sequence information in a messenger RNA.

Nucleotide Synthesis and Salvage

Nucleotides are made from amino acids, carbon dioxide and formic acid in pathways that require large amounts of metabolic energy. Consequently, most organisms have efficient systems to salvage preformed nucleotides. Purines are synthesized as nucleosides (bases attached to ribose). Both adenine and guanine are made from the precursor nucleoside inosine monophosphate, which is synthesized using atoms from the amino acids glycine, glutamine, and aspartic acid, as well as formate transferred from the coenzyme tetrahydrofolate. Pyrimidines, on the other hand, are synthesized from the base orotate, which is formed from glutamine and aspartate.

Xenobiotics and Redox Metabolism

All organisms are constantly exposed to compounds that they cannot use as foods and would be harmful if they accumulated in cells, as they have no metabolic function. These potentially damaging compounds are called xenobiotics. Xenobiotics such as synthetic drugs, natural poisons and antibiotics are detoxified by a set of xenobiotic-metabolizing enzymes. In humans, these include cytochrome P450 oxidases, UDP-glucuronosyltransferasess, and glutathione *S*-transferases. This system of enzymes acts in three stages to firstly oxidize the xenobiotic (phase I) and then conjugate water-soluble groups onto the molecule (phase II).

The modified water-soluble xenobiotic can then be pumped out of cells and in multicellular organisms may be further metabolized before being excreted (phase III). In ecology, these reactions are particularly important in microbial biodegradation of pollutants and the bioremediation of contaminated land and oil spills. Many of these microbial reactions are shared with multicellular organisms, but due to the incredible diversity of types of microbes these organisms are able to deal with a far wider range of xenobiotics than multicellular organisms, and can degrade even persistent organic pollutants such as organochloride compounds.

A related problem for aerobic organisms is oxidative stress. Here, processes including oxidative phosphorylation and the formation of disulfide bonds during protein folding produce reactive oxygen species such as hydrogen peroxide. These damaging oxidants are removed by antioxidant metabolites such as glutathione and enzymes such as catalases and peroxidases.

Thermodynamics of Living Organisms

Living organisms must obey the laws of thermodynamics, which describe the transfer of heat and work. The second law of thermodynamics states that in any closed system, the amount of entropy (disorder) will tend to increase. Although living organisms' amazing complexity appears to contradict this law, life is possible as all organisms are open systems that exchange matter and energy with their surroundings.

Thus living systems are not in equilibrium, but instead are dissipative systems that maintain their state of high complexity by causing a larger increase in the entropy of their environments. The metabolism of a cell achieves this by coupling the spontaneous processes of catabolism to the non-spontaneous processes of anabolism. In thermodynamic terms, metabolism maintains order by creating disorder.

Regulation and Control

As the environments of most organisms are constantly changing, the reactions of metabolism must be finely regulated to maintain a constant set of conditions within cells, a condition called homeostasis. Metabolic regulation also allows organisms to respond to signals and interact actively with their environments. Two closely-linked concepts are important for understanding how metabolic pathways are controlled. Firstly, the *regulation* of an enzyme in a pathway is how its activity is increased and decreased in response to signals. Secondly, the *control* exerted by this enzyme is the effect that these changes in its activity have on the overall rate of the pathway (the flux through the pathway). For example, an enzyme may show large changes in activity (*i.e.* it is highly regulated) but if these changes have little effect on the flux of a metabolic pathway, then this enzyme is not involved in the control of the pathway.

There are multiple levels of metabolic regulation. In intrinsic regulation, the metabolic pathway self-regulates to respond to changes in the levels of substrates or products; for example, a decrease in the amount of product can increase the flux through the pathway to compensate. This type of regulation often involves allosteric regulation of the activities of multiple enzymes in the pathway.

Extrinsic control involves a cell in a multicellular organism changing its metabolism in response to signals from other cells. These signals are usually in the form of soluble messengers such as hormones and growth factors and are detected by specific receptors on the cell surface. These signals are then transmitted inside the cell by second messenger systems that often involved the phosphorylation of proteins.

A very well understood example of extrinsic control is the regulation of glucose metabolism by the hormone insulin. Insulin is produced in response to rises in blood glucose levels. Binding of the hormone to insulin receptors on cells then activates a cascade of protein kinases that cause the cells to take up glucose and convert it into storage molecules such as fatty acids and glycogen. The metabolism of glycogen is controlled by activity of phosphorylase, the enzyme that breaks down glycogen, and glycogen synthase, the enzyme that makes it.

These enzymes are regulated in a reciprocal fashion, with phosphorylation inhibiting glycogen synthase, but activating phosphorylase. Insulin causes glycogen synthesis by activating protein phosphatases and producing a decrease in the phosphorylation of these enzymes.

Evolution

The central pathways of metabolism described above, such as glycolysis and the citric acid cycle, are present in all three domains of living things and were present in the last universal ancestor. This universal ancestral cell was prokaryotic and probably a methanogen that had extensive amino acid, nucleotide, carbohydrate and lipid metabolism. The retention of these ancient pathways during later evolution may be the result of these reactions being an optimal solution to their particular metabolic problems, with pathways such as glycolysis and the citric acid cycle producing their end products highly efficiently and in a minimal number of steps. The first pathways of enzyme-based metabolism may have been parts of purine nucleotide metabolism, with previous metabolic pathways being part of the ancient RNA world.

Many models have been proposed to describe the mechanisms by which novel metabolic pathways evolve. These include the sequential addition of novel enzymes to a short ancestral pathway, the duplication and then divergence of entire pathways as well as the recruitment of pre-existing enzymes and their assembly into a novel reaction pathway. The relative importance of these mechanisms is unclear, but genomic studies have shown that enzymes in a pathway are likely to have a shared ancestry, suggesting that many pathways have evolved in a step-by-step fashion with novel functions being created from pre-existing steps in the pathway.

An alternative model comes from studies that trace the evolution of proteins' structures in metabolic networks, this has suggested that enzymes are pervasively recruited, borrowing enzymes to perform similar functions in different metabolic pathways (evident in the MANET database). These recruitment processes result in an evolutionary enzymatic mosaic. A third possibility is that some parts of metabolism might exist as "modules" that can be reused in different pathways and perform similar functions on different molecules.

As well as the evolution of new metabolic pathways, evolution can also cause the loss of metabolic functions. For example, in some parasites metabolic processes that are not essential for survival are lost and preformed amino acids, nucleotides and carbohydrates may instead be scavenged from the host. Similar reduced metabolic capabilities are seen in endosymbiotic organisms.

Investigation and Manipulation

Classically, metabolism is studied by a reductionist approach that focuses on a single metabolic pathway. Particularly valuable is the use of radioactive tracers at the whole-organism, tissue and cellular levels, which define the paths from precursors to final products by identifying radioactively-labelled intermediates and products. The enzymes that catalyze these chemical reactions can then be purified and their kinetics and responses to inhibitors investigated.

A parallel approach is to identify the small molecules in a cell or tissue; the complete set of these molecules is called the metabolome. Overall, these studies give a good view of the structure and function of simple metabolic pathways, but are inadequate when applied to more complex systems such as the metabolism of a complete cell.

An idea of the complexity of the metabolic networks in cells that contain thousands of different enzymes is given by the figure showing the interactions between just 43 proteins and 40 metabolites to the right: the sequences of genomes provide lists containing anything up to 45,000 genes. However, it is now possible to use this genomic data to reconstruct complete networks of biochemical reactions and produce more holistic mathematical models that may explain and predict their behavior. These models are especially powerful when used to integrate the pathway and metabolite data obtained through classical methods with data on gene expression from proteomic and DNA microarray studies.

A major technological application of this information is metabolic engineering. Here, organisms such as yeast, plants or bacteria are genetically-modified to make them more useful in biotechnology and aid the production of drugs such as antibiotics or industrial chemicals such as 1,3-propanediol and shikimic acid. These genetic modifications usually aim to reduce the amount of energy used to produce the product, increase yields and reduce the production of wastes.

REFERENCES

Dunn L, Rahmanto Y, Richardson D., "Iron uptake and metabolism in the new millennium", *Trends Cell Biology*, 2007.

Bruce Alberts, "Cell Movements and the Shaping of the Vertebrate Body" in *Chapter 21 of Molecular Biology of the Cell*, fourth edition, 2002.

Tichi M, Tabita F., "Interactive control of Rhodobacter capsulatus redox-balancing systems during phototrophic metabolism", *Journal of Bacteriology*, 2001.

11

Cell Biotechnology

Biotechnology is an interdisciplinary field based on a combination of biology and technology to design and produce new molecules, plants, animals, and microorganisms with improved characteristics. Biotechnology offers seemingly unlimited opportunities to combine genes from related or unrelated species to produce useful organisms with desirable properties that were not previously found in nature.

Biotechnology may be thought of as a collection of technologies using animal and/or plant cells, biological molecules, molecular biology processes and genetic engineering for applications in medicine, agriculture and the pharmaceutical industry. The technologies include the use of recombinant DNA for gene cloning and gene transfers between organisms; culture of plant and animal cells and tissues; fusion of animal cells or plant protoplast; the regeneration of whole plants from single cells and the large-scale fermentation processes that use some of these novel organisms for the production of pharmaceuticals, diagnostic tests for disease, feed additives, enzymes and hormones.

ANIMAL CELL BIOTECHNOLOGY

Animal cell biotechnology is an important field in biotechnology. The study of animal cells has helped us gain an insight not only in the structure and function of cells and tissues but also in different physiological, biochemical and immunological processes. Animal cells are known to produce many chemicals of great value to humans. There is an ever-increasing list of biologicals available from animal cells.

Ever since recombinant DNA technology came into being, it was easy to produce these products in bacteria or fungi on a large scale. Large-scale culture of bacteria and fungi is relatively easy. But the main drawbacks of prokaryotes is that they lack post-translational modifications such as phosphorylation, glycosylation, amidation, carboxylation etc. as a result of which the protein is different and may not be functional. Besides, bacteria and yeasts may not excrete their products. Very often, the wild type cells overgrow the recombinants.

Animal cells soon came to be considered as alternatives to bacteria and yeasts for recombinant products. Mammalian cells possess post-translational systems and extraction of products from these cells is possible. As a result, cell lines are being preferred to bacteria and yeasts as cloning systems for developing recombinant products. Different methods of mass cultivation of animal cells have been devised to meet the growing demand for these cells. The choice of the method mainly depends on the cells to be used. Basically, cell lines may be roughly divided into those that grow as a suspension and those that are anchorage dependent.

CULTIVATION OF ANIMAL CELLS

Anchorage Dependent Culture

Anchorage dependent cell lines can be grown by the following methods.

Conventional Methods

These methods are followed routinely in our laboratories.

They include MD bottles, T flasks, Roux bottles and Rollers. If production is to be increased, the number of units has to be increased making the process time consuming and laborious. The result is that the product may not be cost effective. On the other hand, newer trends in manufacture aim at increasing productivity of a single unit.

i) *Multiple surface tissue culture propagator.* This consists of stacks of glass plates maintained in a supporting rack within a glass or acrylic culture vessel. These serve as the substratum on which cells can grow. There is an air-lift pump which circulates the medium at a predetermined rate. A sparger is provided to pass air and CO_2.

ii) *Capillaries.* Semi permeable membranes, in the shape of capillaries 200-300 u in diameter, are bundled together in a glass culture vessel and medium is circulated continuously. The cells grow along the walls of the capillary. The nutrients can diffuse in and products can diffuse out of the semi-permeable membrane. Thus it is possible to operate such a system continuously.

iii) *As aggregates.* This is possible mainly with insect cell lines. Enzyme treatment can release cells growing as a monolayer and they form aggregates ranging from 20-500 u in diameter. These aggregates are then grown as a suspension culture. The cells in the aggregates multiply and release products in the surrounding medium.

iv) *Encapsulation.* Cells can be trapped inside beads made up of agarose, karageenan, calcium alginate etc. These beads are then introduced into stirred reactors. The products released remain localised in the material which goes to make the beads thereby making downstream processing very easy.

v) *Microcarriers.* These are bead-like structures ranging from 100-180 u in diameter that can be held in homogeneous suspension in stirred reactors. They are capable of providing a large surface area to cells in small volumes of media. Commercially available microcarriers have been made from a variety of materials such as DEAE Sephadex, cellulose, glass, gelatin etc. Cells attach and spread on the microcarriers and gradually grow into confluent monolayers. The microcarriers are held in a medium that is constantly stirred. The features of both suspension and monolayer culture are brought together in this system. Porous microcarriers, which allow growth of cells inside them, are termed as macrocarriers. Macrocarriers are also available commercially eg. Cultispher-G. These type of microcarriers will produce better yields, the cells growing in them are not exposed to shearing forces in fermentors.

vi) *Packed bed reactors*. Microcarriers, macrocarriers or encapsulated beads could be used in fixed-bed reactors. The cells are immobilised in a matrix and the culture fluid is circulated in a closed loop. There is no agitation system. If the bed of immobilised cells is disturbed by the circulating medium it is said to be a fluidised bed reactor. Such a process achieves a high degree of aeration and agitation.

Suspension Culture

The following methods are available for growing cells as a suspension.

i) *Bioreactors*. These are essentially the classical fermentors used for the cultivation of bacterial and fungal cultures. They consist of a glass or metal vessel covered by a head plate. There is an agitator shaft with impellers, which stir the culture. Different probes are introduced into the vessel and these monitor process parameters like pH, temperature, dissolved Oxygen and CO_2 etc. Fermentors can be scaled up to hundreds of liters. Largest reported fermentor is of 8000 liters.

ii) *Hollow fiber reactor*. These consist of semi-permeable membranes packed together. In the lumen of these "fibers", cells can grow. The semi-permeable membrane has a definite cut-off and hence nutrients can diffuse into the lumen and products and by-products of metabolism can diffuse out. Since the medium is constantly circulated, products can be obtained continuously. Large surface area to volume is provided for cell growth.

iii) *Air-lift fermentors*. These have a cylinder inside the culture vessel through which air is passed for circulating the medium. It is a very well aerated system. This type of fermentor is used mainly for monoclonal antibody production.

iv) *Chemostats*. This is a continuously stirred tank reactor with an inlet and outlet so that medium can be passed continuously at a fixed rate keeping the volume constant. The theory for chemostats was established by Monod and Novick and Szillard independently in 1942. The culture is first grown in batch mode and when cells reach the log phase they are grown in continuous mode. The medium employed has one growth-limiting nutrient and the rest are in excess. Hence cell growth is proportional to the growth-limiting nutrient. Steady state is obtained by a continuous and constant feed and harvesting rate where the dilution rate is determined by growth rate.

Modes of Suspension Culture Technology

The following modes of cultivation are available for cells grown as a suspension in bioreactors.

i) *Batch culture*. Until a few years ago, most biologicals were produced as a single batch. Growth is determined by the rate of nutrient depletion and toxic metabolite accumulation. Vaccines and interferons are still produced in this mode. The yields are low compared to other modes. In this type of culture, cells are inoculated into a fixed volume of medium. They consume nutrients as they grow and release toxic metabolites into the surrounding medium. The environment changes continuously. Eventually, cell multiplication ceases due to accumulation of toxic metabolites and depletion of nutrients.

ii) *Fed-batch culture.* The culture is fed intermittently either with the limiting nutrient or with the whole growth medium. Cell growth is limited not by nutrients but by accumulation of waste products. In this mode, a higher cell concentration is achieved and cell viability is maintained for larger periods compared to the batch mode.

iii) *Continuous culture.* This mode is based on the principle established by Monod Homoestatic culture conditions are maintained in chemostats with no fluctuations of nutrients, metabolites or cell numbers. Once established they can be maintained at high cell density and high product yield for a long period.

iv) *Perfusion culture.* In contrast to continuous culture where steady state is maintained by constant dilution of the culture, in perfusion cultures, the cells are physically retained in the vessel. Spent medium is withdrawn continuously from the culture system and an equal volume of fresh medium is added. Cell concentration increases constantly until it becomes limited at high cell population densities.

Fermentation Technology

Fermentors have been used for the cultivation of bacteria and yeasts for a long time. Initially, fermentation was synonymous with alcohol production. Later, bacteriologists learnt to use the same principles for the preparation of vitamins, organic acids, antibiotics etc. This led to the rapid development of a variety of fermentors and methodologies.

It is but natural that the same principles be applied for the mass cultivation of animal and plant cells. However, adaptation of these processes is required. Cultivation of plant and animal cells is difficult, mainly due to the slow metabolism of these cells which is reflected by slow cell growth. Animal cells have complex nutritional requirements as compared to bacteria and yeasts. Thus, the media are complex and costly and prone to contamination. Animal cells lack the cell wall present in bacterial cells making them fragile and shear sensitive. Therefore, the agitation and aeration systems have to be designed differently.

Cell densities achieved are low resulting in low concentration of products. Downstream processing procedures required to concentrate and purify the product lead to increased costs. Inspite of these drawbacks, fermentors have been used for growing animal cells for the last few decades. Different cell lines like BHK-21, LS, Namalwa cells etc have been grown in fermentors as submerged cultures for the production of viral vaccines and other products. The shear sensitivity and fragility of animal cells are overcome by the introduction of novel impellers, which are paddle shaped.

Direct gas sparging produces bubbles capable of rupturing cells and hence supply of gases needs to be done via diffusion through a silicone tubing. The medium contains plenty of serum proteins capable of generating froth. Therefore, agitation has to be slow and gentle. For high-density cultivation, oxygen supply becomes critical. The silicone tubing method of aeration has many advantages. No bubbles are formed and the oxygen transfer rate has been found to be satisfactory. However, the winding of the fragile tubing is difficult and is limited to small scale reactors used in laboratories. Thus suitably modified fermentors can be used for the mass culture of animal cells growing as suspension cultures. If an anchorage-dependent cell line needs to be grown, it becomes necessary to use a carrier system such as microcarriers.

DEVELOPMENT OF MICROCARRIERS

With the development of microcarriers in 1967 by van Wezel the production of biologicals from anchorage dependent cells is a dream come true. van Wezel made microcarriers composed of cross-linked dextran beads charged with tertiary amine groups (DEAE) having an exchange capacity of 3.5 milli equivalent/gm dry weight. He demonstrated the growth of primary cells and cells from a human diploid cell strain on them as well as the propagation of poliomyelitis virus in the cells grown on microcarriers.

These microcarriers faced the problem of cell attachment due to large charge densities on their surfaces. It was observed that above a bead concentration of 1 gm/liter, there were increased innoculum losses, long lag periods and diminished capacity for cell growth. Complete cell death was seen at bead concentrations of more than 2 gm/liter. To overcome this problem, many approaches were tried like coating the beads with serum proteins, nitrocellulose or carboxymethyl cellulose.

It was Levine et al 1979 who developed microcarriers of reduced ion exchange capacity (1.5 meq./gm dry weight). These could be used at concentrations as high as 6 gms/liter with no deleterious effect on cell growth. These low charged beads could be used to cultivate primary, diploid as well as established cell lines. Subsequently a lot of work was done to improve the quality of the beads. Different types of microcarriers were developed and patented.

Henderson T. M has obtained a patent for glass microcarriers. These have an outer layer of glass below which a layer of magnetic material has been added so that the microcarriers may be separated from the culture medium under the influence of a magnetic field. Cross-linked dextran has been found to be the best material for making microcarriers. Also, microcarriers were derived such that the surface charge was limited to the surface layers only. In such microcarriers the absorption of proteins like IgG or albumin present in the serum was low. Collagen has also been used for producing microcarriers. These beads have the advantage that they can be digested by enzymes, thereby facilitating harvesting.

Microcarrier Cultures

This system has the following advantages over other methods of large-scale cultivation:

- High surface area to volume ratio can be achieved which, can be varied by changing the microcarrier concentration. This leads to high cell densities per unit volume with a potential for obtaining highly concentrated cell products.
- Cell propagation can be carried out in a single high productivity vessel instead of using many low productivity units, thus achieving a better utilisation and a considerable saving of medium.
- Since the microcarrier culture is well mixed, it is easy to monitor and control different environmental conditions such as pH, pO_2, pCO_2 etc.
- Cell sampling is easy.
- Since the beads settle down easily, cell harvesting and downstream processing of products is easy.
- Microcarrier cultures can be relatively easily scaled up using conventional equipment like fermenters that have been suitably modified.

Because of the many advantages of the technique itself, it has gained great popularity. Thus, a large variety of microcarriers are available in the market. Due to the growing need for large quantities of anchorage-dependent cells, a variety of microcarriers are now available in the market. The availability of a wide range of microcarrier products in the market makes it necessary to define the exact requirements of a good microcarrier.

Requirements for an Optimum Microcarrier

— The surface properties of the beads should be such that cells can adhere and proliferate rapidly. In other words, the contour should be even.

— The density of the beads should be slightly more than that of the culture medium, so as to facilitate easy separation. Conventional culture media are aqueous in nature and have densities ranging from 1.03-1.09 g/cc. However the density should not exceed a certain limit the optimum range being 1.03-1.045 gms/ml. Gentle stirring, which will not harm the shear-sensitive cells, should be sufficient to keep them in suspension. If the beads settle down cell growth will be prevented.

— The size-distribution of the beads should be narrow so that an even suspension of all microcarriers is achieved and cells attain confluency at approximately the same time.

— The optical properties should enable easy microscopic observation.

— They should be non-toxic not only for the survival and good growth of the cells but also for cell culture products that are used for veterinary or clinical purposes.

— The matrix of the beads should be such that collisions, which occur during stirring of the culture, do not cause fragmentation of the beads.

— Ideally, the beads should be such that they can be reused.

Choice of Microcarriers

The choice of the microcarrier depends primarily on the objective of the culture. For instance, in the case of extracellular viruses and cell products obtained from established cell lines, dextran microcarriers prove to be the best. If the cells need to be trypsinised or detached then glass or gelatin beads are preferred. When a particular type of cell tends to have a low plating efficiency, it is important to use that microcarrier which enables attachment of the maximum number of cells. The choice of the microcarrier also depends on the morphology of the cells. For instance, when culturing cells with fibroblast-like morphology, microcarriers with a layer of surface charge on the periphery of the bead tends to show better attachment of cells as compared with others. For cells with an epithelial-like morphology, the attachment is better on beads coated with a protein like gelatin. Macrocarriers can be used when cells are extremely shear sensitive.

Adhesion of Cells to Microcarrier Surface

The surface on which the cells are grown as well as the cells at physiological pH have either a net negative or a positive charge. The charge density on these surfaces rather than their polarity is responsible for attachment and spreading of cells. It is also known that cell adhesion is mediated by specific cell surface receptors.

Maroudas's theory of attachment states that cell surface contact is bridged by amphoteric proteins. According to him, the prerequisite for attachment of cells onto any surface is the adsorption of protein factors to culture surface after which the contact between cells and surface can take place. The cells attach onto the substratum before which they produce their own glycoproteins and matrix proteins and then spread out). It is this matrix that adheres to the charged surface and the cells then bind to the matrix via specific receptors. Therefore many a times it has been observed that surfaces that have been conditioned by previous cell growth often provide a better surface for attachment.

Three major classes of transmembrane proteins have been shown to be involved in cell-cell and cell-substrate adhesion viz.

i) Cell-cell adhesion molecules (CAMS) which are calcium independent and Cadherins which are calcium dependent

ii) Integrins which are receptors for matrix molecules such as fibronectin, laminin and collagen which bind to them with the help of a arginine-glycine-aspartic acid sequence known as the RGD motif

iii) Transmembrane proteoglycans which interact with matrix molecules like collagen but not with the help of the RGD motif.

Cell adhesion also depends on a functional contractile system. It involves multiple contacts with the surface. Next numerous filopodia are formed. They fit into a lattice structure formed by the glycoproteins on the substratum. This is followed by cytoplasmic webbing and flattening of the cell mass. Thus the cells get attached onto the substratum. Once the cells have attached onto the microcarriers, they grow using the nutrients provided in their culture medium. The complexities of this culture system are mainly due to the large number of parameters affecting cell yield. Medium composition now assumes importance since it contains the carbon and nitrogen source, the energy source, growth factors and dissolved oxygen and other gases. Besides nutrient limitation, growth of cells is also affected by the accumulation of toxic metabolites. Other important considerations are environmental factors like pH and temperature and shear sensitivity of the cells, especially in case of microcarrier cultures employing spinner bottles and fermentors.

Applications of Microcarrier Cultures

A wide range of cells have been cultured on microcarriers. For instance, cells from invertebrates, from fish, birds and cells of mammalian origin have been cultivated on microcarriers. Transformed and normal cell lines, fibroblastic and epithelial cells and even genetically engineered cells have been cultivated on microcarriers for various biologicals such as for the production of immunologicals like interferons, interleukins, growth factors etc. Cells cultured on microcarriers also serve as hosts for a variety of viruses that are used as vaccines like foot and mouth disease or rabies.

Microcarrier cultures have found a wide number of applications other than mass cultivation as well. Cells growing on microcarriers serve as an excellent tool for studying different aspects of cell biology such as cell-to-cell or cell-to-substratum interactions. Cell differentiation and maturation, metabolic studies have also been carried using microcarriers. Such cells can also be used for electron microscopic examinations or for the isolation of cell organelles such as the cell membrane. Also, this system is essentially a three-dimensional system and serves as a good 3-D model. Similarly, co-cultivation of cells can be done using this system. Microcarriers have also been used for the depletion of macrophages from a population of spleen cells.

DEAE-dextran microcarriers can potentiate stimulation of lymphocytes by concanavalin A (con A). Microcarrier beads confluent with allogenic tumour cells can be injected in mice to increase humoral and cell-mediated immunity. Plant protoplasts can be immobilised on DEAE-dextran microcarriers. Due to the large surface area to volume ratio provided by microcarriers, they are now successfully being used for a variety of biologicals on a laboratory as well as an industrial scale of even 4000 liters.

Mammalian Cell Culture

Mass culture of animal cell lines is fundamental to the manufacture of viral vaccines and many products of biotechnology. Biologicals produced by recombinant DNA (rDNA) technology in animal cell cultures include enzymes, hormones, immunobiologicals (monoclonal antibodies, interleukins, lymphokines), and anticancer agents. Although many simpler proteins can be produced using rDNA in bacterial cultures, more complex proteins that are glycosylated (carbohydrate-modified), currently must be made in animal cells.

An important example of such a complex protein is the hormone erythropoietin. The cost of growing mammalian cell cultures is high, so research is underway to produce such complex proteins in insect cells or in higher plants. Vaccines for polio, measles, mumps, rubella, and chickenpox are currently made in cell cultures. Due to the H5N1 pandemic threat, research into using cell culture for flu vaccines is being funded by the United States government. Novel ideas in the field include recombinant DNA-based vaccines, such as one made using human adenovirus (a common cold virus) as a vector, or the use of adjuvants.

Cells can be isolated from tissues for ex vivo culture in several ways. Cells can be easily purified from blood, however only the white cells are capable of growth in culture. Mononuclear cells can be released from soft tissues by enzymatic digestion with enzymes such as collagenase, trypsin, or pronase, which break down the extracellular matrix. Alternatively, pieces of tissue can be placed in growth media, and the cells that grow out are available for culture. This method is known as explant culture.

Cells that are cultured directly from an animal or person are known as primary cells. With the exception of some derived from tumours, most primary cell cultures have limited lifespan. After a certain number of population doublings cells undergo the process of senescence and stop dividing, while generally retaining viability. An established or immortalised or transformed cell line has acquired the ability to proliferate indefinitely either through random mutation or deliberate modification, such as artificial expression of the telomerase gene. There are numerous well established cell lines representative of particular cell types.

Maintaining Cells in Culture

Cells are grown and maintained at an appropriate temperature and gas mixture in a cell incubator. Culture conditions vary widely for each cell type, and variation of conditions for a particular cell type can result in different phenotypes being expressed. Varied factor in culture systems is the growth medium. Recipes for growth media can vary in pH, glucose concentration, growth factors, and the presence of other nutrient components. The growth factors used to supplement media are often derived from animal blood, such as calf serum. These blood-derived ingredients pose the potential for contamination of derived pharmaceutical products with viruses or prions. Current practice is to minimise or eliminate the use of these ingredients where possible. Some cells naturally live without attaching to a surface, such as cells that exist in the bloodstream. Others require a surface, such as most cells derived from solid tissues. Cells grown unattached to a surface are referred to as suspension cultures. Other adherent cultures cells can be grown on tissue culture plastic, which may be coated with extracellular matrix components to increase its adhesion properties and provide other signals needed for growth.

Manipulation of Cultured Cells

As cells generally continue to divide in culture, they

generally grow to fill the available area or volume. This can generate several issues:

— Nutrient depletion in the growth media
— Accumulation of apoptotic/necrotic (dead) cells
— Cell-to-cell contact can stimulate cell cycle arrest, causing cells to stop dividing
— Cell-to-cell contact can stimulate promicuous and unwanted cellular differentiation

These issues can be dealt with using tissue culture methods that rely on sterile technique. These methods aim to avoid contamination with bacteria or yeast that will compete with mammalian cells for nutrients and/or cause cell infection and cell death. Manipulations are typically carried out in a biosafety hood or laminar flow cabinet to exclude contaminating micro-organisms. Antibiotics can also be added to the growth media. Amongst the common manipulations carried out on culture cells are media changes, passaging cells, and transfecting cells.

Media Changes

The purpose of media changes is to replenish nutrients and avoid the build up of potentially harmful metabolic biproducts and dead cells. In the case of suspension cultures, cells can be separated from the media by centrifugation and resuspended in fresh media. In the case of adherent cultures, the media can be removed directly by aspiration and replaced.

Passaging Cells

Passaging or splitting cells involves transferring a small number of cells into a new vessel. Cells can be cultured for a longer time if they are split regularly, as it avoids the senescence associated with prolonged high cell density. Suspension cultures are easily passaged with a small amount of culture containing a few cells diluted in a larger volume of fresh media. For adherent cultures, cells first need to be detached; this was historically done with a mixture of trypsin-EDTA, however other enzyme mixes are now available for this purpose. A small number of detached cells can then be used to seed a new culture.

Transfection

Transfection is the introduction of foreign DNA into eukaryotic or prokaryotic cells, such as animal or bacterial cells. Transfection typically involves opening transient "holes" or gates in cells to allow the entry of extracellular molecules, typically supercoiled plasmid DNA, but also siRNA, among others. Cells that have been manipulated to accept foreign DNA (cells with "holes") are called "competent cells". Transfection differs from transformation since the DNA is not generally incorporated into the cell's genome, it is only transiently expressed.

There are various methods of introducing foreign DNA into a cell. One of the cheapest (and least reliable) ones is transfection by calcium phosphate, originally discovered by S. Bacchetti and F. L. Graham in 1977. HEPES-buffered saline solution (HeBS) containing phosphate ions is combined with a calcium chloride solution containing the DNA to be transfected. When the two are combined, a fine precipitate of calcium phosphate will form, binding the DNA to be transfected on its surface. The suspension of the precipitate is then added to the cells to be transfected (usually a cell culture grown in a monolayer). By a process not entirely understood, the cells take up some of the precipitate, and with it, the DNA. Other methods of transfection include electroporation, heat shock, and proprietary transfection reagents such as Fugene.

Other methods use highly branched organic compounds, so-called dendrimers, to bind the DNA and get it into the cell. A very efficient method is the inclusion of the DNA to be transfected in liposomes, i.e. small, membrane-bounded bodies that are in some ways similar to the structure of a cell and can actually fuse with the cell membrane, releasing the DNA into the cell. For eukaryotic cells, lipid- cation based transfection is more typically used, because the cells are more sensitive.

A direct approach to transfection is the gene gun, where the DNA is coupled to a nanoparticle of an inert solid (commonly gold) which is then "shot" directly into the target cell's nucleus. DNA can also be introduced into cells using viruses as a carrier. In such cases, the technique is called viral transduction, and, the cells are said to be transduced.

For most applications of transfection, it is sufficient if the transfected gene is only transiently expressed. Since the DNA introduced in the transfection process is usually not inserted into the nuclear genome, the foreign DNA is lost at the latest when the cells undergo mitosis. If it is desired that the transfected gene actually remains in the genome of the cell and its daughter cells, a stable transfection must occur. To accomplish this, another gene is co-transfected, which gives the cell some selection advantage, such as resistance towards a certain toxin. Some of the transfected cells will, by chance, have inserted the foreign genetic material into their genome. If the toxin, towards which the co-transfected gene offers resistance, is then added to the cell culture, only those few cells with the foreign genes inserted into their genome will be able to proliferate, while other cells will die. After applying this selection pressure for some time, only the cells with a stable transfection remain and can be cultivated further. A common agent for stable transfection is Geneticin, also known as G418, which is a toxin that can be neutralised by the product of the neomycin resistant gene.

Human Cell Lines

Cell lines that originate with humans have been somewhat

controversial in bioethics, as they may outlive their parent organism and later be used in the discovery of lucrative medical treatments. In the pioneering decision in this area, the Supreme Court of California held in 1990 that human patients have no property rights in cell lines derived from organs removed with their consent. It is estimated that about 20% of human cell lines are not the kind of cells they were generally assumed to be. The reason for this is that some cell lines exhibit vigorous growth and thus can cross-contaminate cultures of other cell lines, in time overgrowing and displacing the original cells. The most common contaminant is the HeLa cell line. While this may not be of significance when general properties such as cell metabolism are researched, it is highly relevant e.g. in medical research focusing on a specific type of cell. Results of such research will be at least flawed, if not outright wrong in their conclusion, with possible consequences if therapeutic approaches are developed based on it.

Hybridoma cells

Hybridoma cells are cells that have been engineered to produce a desired antibody in large amounts. To produce monoclonal antibodies, B-cells are removed from the spleen of an animal that has been challenged with the relevant antigen. These B-cells are then fused with myeloma tumor cells that can grow indefinitely in culture (myeloma is a B-cell cancer). This fusion is performed by making the cell membranes more permeable. The fused hybrid cells (called hybridomas), being cancer cells, will multiply rapidly and indefinitely and will produce large amounts of the desired antibodies. They have to be selected and subsequently cloned by limiting dilution. Supplemental media containing Interleukin-6 (such as briclone) are essential for this step.

Selection occurs via culturing the newly fused primary hybridoma cells in selective-media, specifically media containing 1x concentration HAT for roughly 10-14 days. After using HAT it is often desirable to use HT containing media. Cloning occurs after identification of positive primary hybridoma cells. Clone by limited dillution. While some may believe that IL-6 is essential for this step, it is not necessary to add that expensive supplement, rather use 50% heat-inactivated FBS for the first week. Add 10% FBS DMEM to the clone culture plate after screening for single colony wells.

Biotechnology and Animal Medicines

Biotechnology has yielded new and improved medicines for animals that help lower production costs and improve animal well being by fighting diseases caused by bacteria and parasites. For example, scientists have identified a new anti-bacterial compound that may serve as a substitute for using some antibiotics in animals, a practice that has been criticised for contributing to the increased prevalence of drug-resistant bacteria in human infections. New or enhanced animal vaccines have also been developed through modern biotechnology techniques. Vaccines are now used to prevent diseases including: foot and mouth disease, scours, brucellosis, shipping fever, feline leukemia, rabies, and infections affecting cultivated fish. Biotechnology has led to the development of rapid test kits to diagnose the health of livestock and companion animals.

Biotechnology and Transgenic Animals

Advances in scientific discovery and laboratory techniques in the last half of the twentieth century resulted in the ability to manipulate the deoxyribonucleic acid (DNA) of organisms and gave rise to transgenic animals. The use of transgenic animals may accelerate classical breeding programs and provide a means for the economical production of life-saving pharmaceuticals. To understand how a transgenic animal is produced, it is necessary to review the basic components and functions of living organisms.

Genes and the Genome

Animals are made of billions of cells all working together. Every cell of the animal has a complete "instruction manual" or genome that is inherited from the parents of the animal as a combination of their genomes. The genome resides in the nucleus of the cell. Genes are found within the genome and serve as the "words" of the instruction manual. When a cell reads a word, a specific protein is produced. Proteins give an individual cell, and therefore the animal, its form and function. Genes are written using the four-letter alphabet A, C, G, T. The letters stand for four chemicals called bases, which together compose DNA.

DNA is universal in nature, meaning that the four chemical bases of DNA are the same in all living organisms. Consequently, a gene from one organism can function in another organism. When a new gene is put into an animal's genome it is said to be transgenic. Scientists have also discovered how to genetically modify when and/or where in an animal a gene is expressed.

Making a Transgenic Animal

One way to produce transgenic animals is through a technique called microinjection. Once scientists have identified and isolated the piece of DNA comprising the gene to be transferred, it is injected into a fertilised egg of the desired animal using a very small glass needle visualised under a microscope. In approximately one percent of the injected eggs, the gene becomes a new "word" in the egg's "instruction manual" by physically combining with the egg's genome. Ideally, the new gene integrates into the genome before the egg begins to divide. If this occurs, every cell in the animal can contain the new protein and the animal will pass the gene on to its offspring.

After injection of the gene, the fertilised egg is implanted into a surrogate mother where it fully develops into a transgenic animal.

Traits Being Introduced Into Animals

Currently, the only routine commercial use of transgenic animals (primarily mice) is in the area of human disease research. One way to characterise the range of genetic modifications that are being considered for use in animals is in the three broad areas of input, output, and value-added traits. Examples of each are described below.

Input traits

An "input" trait helps livestock and dairy producers by increasing production efficiency. Input traits that are being investigated for use in animals:

— Faster, more efficient growth rates

— Increased production of milk or wool

— Resistance to diseases caused by viruses and bacteria

Output traits

An output trait helps consumers or downstream processors by enhancing the quality of the animal product. Output traits that may prove to be beneficial:

— Leaner, more tender beef and pork

— Milk that lacks allergenic proteins, or results in increased amounts of cheese and yogurt.

Value-added traits

By adding or modifying genes, animals can function in completely new ways.

— Producing large amounts of therapeutic proteins in animal milk may be an efficient, relatively low cost method to manufacture many proteins used to treat human diseases or proteins that have industrial value.

— Transplanting animal organs into humans, or xenotransplantation, can be made more successful by genetically modifying the organs so that they are not as readily rejected by the human immune system.
— Development of animals that serve as models for human diseases to help scientists better understand prevention and treatment strategies.

Transgenic Animal Biotechnology

One concern of transgenic animal technology is the welfare of the animals. Developmental and health abnormalitites have been reported in conjunction with its use; therefore, researchers must take care to minimise animal suffering. The inadvertant release or escape of transgenic animals (particularly fish) into the wild where they could breed or compete with the natural population is often cited as a potential risk to the environment. The actual risk associated with this will depend on the type of animal and the nature of the genetic modification; however, where appropriate, procedures must be in place to alleviate this concern.

Transgenic Fish and Shellfish

Aquaculture, worth an estimated $46 billion in 1997, produces approximately one-third of all fish and shellfish consumed by humans. Researchers are working with transgenic catfish, carp, tilapia, striped bass, clams, oysters, shrimp, and abalone.

Traits being tested in transgenic fish include:
— Growth rates that are 3-11 times faster with more efficient feed utilisation
— Increased tolerance to cold water
— Improved disease resistance

Biotechnology and Cloning Animals

Advances in biotechnology have allowed scientists to make

genetically identical copies or clones of animals. Duplication of an organism's genome occurs naturally when identical twins are born or when a plant is grown from a cutting of another plant. However, the world really took notice of cloning in 1997 when a group of Scottish researchers announced the birth of Dolly the sheep, which had been cloned using a single cell from an adult sheep. Dolly had only one "parent;" her nuclear genome was exactly like her "mother's" instead of being a combination of two parents. Therefore, Dolly could generally be thought of as her mother's identical twin. To produce Dolly, scientists took an egg from a sheep and removed its nucleus, rendering it unable to function or develop. Next, they took a cell with an intact genome from a different adult sheep and fused it to the sheep egg which lacked a genome. The egg, with its new genome, was stimulated to begin developing into an embryo and was implanted into a surrogate sheep where it grew normally, resulting in the birth of Dolly. Dolly later gave birth to normal lambs.

Benefits and Risks of Cloning

Researchers have cloned other mammals including cows, goats, pigs, and mice. However, the overall low rate of successful cloning and frequent occurrence of developmental abnormalities in cloned animals demonstrate the need for further research before cloning will be practical. It has also been reported that cloned animals may exhibit health problems throughout their life. Cloned animals may age prematurely as Dolly was diagnosed with arthritis at a seemingly young age and cloned mice had a shorter than normal life span. Additionally, it was demonstrated that cloned mice were both larger in size and heavier than a control group of non-cloned mice.

If advances in animal cloning technology were to overcome the current obstacles, the most obvious benefit would be the ability of a farmer to have a herd of superior performing animals in one generation. Breeding companies could sell cloned embryos in a manner similar to the way in which semen is currently marketed. A potential drawback of this practice would be the loss of genetic diversity in livestock herds, but this could be avoided by limiting the number of cloned embryos of a given animal that were sold. It has also been proposed that cloning could be used to increase the population of animals in an endangered species. The mouflon sheep, which is a wild Mediterranean sheep with less than 1000 animals remaining, was successfully cloned. Additionally, scientists are attempting to clone an endangered wild Asian ox, called the guar and possibly the giant Panda. Although possible, a recovering population of cloned animals would be hindered by a lack of genetic diversity and would not address the larger issue of how the animal became endangered.

Mapping of Animal Genes

Biotechnology may be defined as the application of our advancing understanding of living organisms and their components to create industrial products and processes. All animal breeding, either from farm livestock such as dairy or beef cattle, or for companion animals such as the many different breeds of dogs, is a type of biotechnology that has been going on for many centuries. Breeders have selected animals that show particular characteristics or traits and used them in breeding programmes to ensure that these traits are retained. What they have been doing is selecting animals with particular combinations of genes. We now know that this type of conventional breeding of animals involves swapping hundreds of genes, most of which are unidentified.

Modern molecular biology is making it possible to identify how different genes control different characteristics. Scientists are "mapping" the genes on the chromosomes, to that they can see where genes are located and the extent to which they tend to be co-inherited. This will help identify which combinations give the traits that are desired.

Some traits are determined by a single gene and these are relatively easy to study. But others, including commercially important traits such as growth rate and feed conversion efficiency, are controlled by many genes working together. One important research goal is to find genes that make animals resistant to diseases. Selecting for the genes will help produce healthier animals and reduce farmers' reliance on antibiotics and other veterinary drugs. There are several important differences between traditional breeding and genetic modification

— *Traditional breeding*. Large numbers of unknown genes are swapped in crosses Undesirable genes may be inherited together with desirable ones. Crosses can be made only between individuals of the same species. Progress is slow because the progeny must be grown and those showing the desired traits selected for further breeding.

— *Genetic modification*. Only individual, or a small number, of well characterised genes are transferred. Genes can be introduced from any species. Because the individual genes can be labelled their presence can be detected very rapidly, enabling much faster selection. As well as improving livestock for traditional agricultural purposes, genes can be introduced to enable the animals to make other products, for example, medicines.

As well as identifying and isolating individual genes, molecular biology enables scientists to transfer these genes from one individual to another—so called genetic modification or transgenics. This could further accelerate livestock improvement but its usefulness in the foreseeable future is most likely in the area of obtaining high value medical products from animals rather than in agriculture.

There are five main areas in which the production of transgenic animals offer potential benefits. To improve animal health—long term it should be possible to identify genes that confer resistance to disease and to introduce these into livestock, so reducing reliance on antibiotics and chemicals to control disease.

To produce therapeutics for human medicine—recently a transgenic cockerel has been bred by introducing genes into the newly fertilised egg yolk. This offers the potential for producing high value therapeutics in the eggs of genetically modified chickens. To provide models of human disease—the function and control of genes implicated in human disease can be studied in animals. To provide tissue for use in human transplant surgery—the recognition molecules on animal cells may be tailored so that they are no longer rejected as foreign by humans. To improve the efficiency of livestock breeding and to increase the range and quality of livestock products.

Newly fertilised egg cells contain two pronuclei (one originating from the egg cell and one from the sperm) each of which carries half the full complement of chromosomes. The pronuclei fuse to form the nucleus of the fertilised cell which then divides to form the embryo.

Newly fertilised cells can be isolated and immobilised under a microscope. A very fine glass needle is used to inject DNA directly into one of the pronuclei. The egg is then transferred into a foster mother where it grows and divides normally to produce the embryo. A new technique called nuclear transfer enables scientists to remove the nucleus from an egg cell and replace it with a nucleus from another cell. In this way the complete genetic information of the donor cell is incorporated into the recipient.

REFERENCES

Griffiths J.B., "Cell Products : An overview", Ed. R.E. Spier and J.B.Griffiths, *Animal Cell Biotechnology*, Academic press, 1985.

Spier, R., E. and Horaud, F., "The biotechnological future for animal cells in cultures", Animal Cell Biotechnololgy, 1985.

Spier R. E. and Fowler M.W., "Animal and plant cell culture", In *Comprehencive Biotechnology*, 1985.

van Wezel A.L., "Growth of cell strains and primary cells on microcarriers in homogeneous cultures", *Nature*, 1967.

Bibliography

Alberts B, Johnson A, Lewis J, Raff M, Roberts K, Walter P., "Chapter 5: DNA Replication, Repair, and Recombination", *Molecular Biology of the Cell*, Garland Science, 2002.

Alberts B. et al., *Molecular Biology of the Cell*, 2d ed. 1989.

Albrecht-Buehler, G., "Does the geomentric design of centrioles imply their function?", *Cell Motility*, 1981

Ambrose, E. J., and Easty, D. M., *Cell Biology: Reading*, Mass., Addison-Wesley, 1970.

Baker, J. J. W., and Allen, G. E., *The Study of Biology* [2nd ed.], Reading, Mass., Addison-Wesley, 1971.

Bakkenist C.J, Kastan M.B., "DNA damage activates ATM through intermolecular autophosphorylation and dimer dissociation", *Nature*, Jan 30 2003.

Berg J, Tymoczko J.L., Stryer L., *Biochemistry*, 6th ed., San Francisco: W. H. Freeman, 2006.

Berg J.M, Tymoczko J.L, Stryer L, Clarke N.D., "Chapter 27, Section 4: DNA Replication of Both Strands Proceeds Rapidly from Specific Start Sites", *Biochemistry*, W.H. Freeman and Company, 2002.

Bruce Alberts, "Cell Movements and the Shaping of the Vertebrate Body" in *Chapter 21 of Molecular Biology of the Cell*, fourth edition, 2002.

Clarke M.B. and V. Sperandio, "Events at the host-microbial interface of the gastrointestinal tract III", Cell-to-cell signalling among microbial flora, host, and pathogens: there is a whole lot of talking going on" in *American journal of physiology, Gastrointestinal and liver physiology*, 2005.

Dowben, R. M., *Cell Biology*, New York, Harper & Row, 1971.

Dunn L, Rahmanto Y, Richardson D., "Iron uptake and metabolism in the new millennium", *Trends Cell Biology*, 2007.

Friedberg E.C, Walker G.C, Siede W, Wood R.D., Schultz R.A., Ellenberger T., *DNA Repair and Mutagenesis, part 3*, ASM Press. 2nd ed. 2006.

Griffiths J.B., "Cell Products : An overview", Ed. R.E. Spier and J.B.Griffiths, *Animal Cell Biotechnology*, Academic.press, 1985.

Jun Ito and Hiroo Fukuda, "ZEN1 Is a Key Enzyme in the Degradation of Nuclear DNA during Programmed Cell Death of Tracheary Elements", *The Plant Cell*, December 2002.

Lackie J. M. and J. A. Dowe, ed., *The Dictionary of Cell Biology*, 1989.

Lodish H, Berk A, Matsudaira P, Kaiser CA, Krieger M, Scott MP, Zipursky SL, Darnell J., *Molecular Cell Biology*, WH Freeman: New York, NY. 5th ed., 2004.

Mazal Solomon, et al., "The Involvement of Cysteine Proteases and Protease Inhibitor Genes in the Regulation of Programmed Cell Death in Plants", *The Plant Cell*, Vol. March 1999.

Mohamed O. A. , M. Jonnaert, C. Labelle-Dumais, K. Kuroda, H. J. Clarke and D. Dufort, "Uterine Wnt/beta-catenin signalling is required for implantation", in *Proceedings of the National Academy of Sciences of the United States of America Volume*, 2005.

Mohamed Ouhammouch, Robert E. Dewhurst, Winfried Hausner, Michael Thomm, and E. Peter Geiduschek, "Activation of archaeal transcription by recruitment of the TATA-binding protein", *Proceedings of the National Academy of Sciences of the United States of America*, 2003

Morgan D.O., *The Cell Cycle: Principles of Control*, New Science Press: London, 2007.

Prescott D. M., *Cells*, 1988.

Richard A. Lockshin, Carroll M. Williams, "Programmed cell death—II. Endocrine potentiation of the breakdown of the intersegmental muscles of silkmoths", *Journal of Insect Physiology*, 1964.

Robert J., *Brooker Genetics: analysis and principles*, 2nd edition, New York: McGraw-Hill, 2005.

Spier R. E. and Fowler M.W., "Animal and plant cell culture", In *Comprehencive Biotechnology*, 1985.

Spier, R., E. and Horaud, F., "The biotechnological future for animal cells in cultures", Animal Cell Biotechnololgy, 1985.

Thomas L., *The Lives of a Cell*, 1974.

Tichi M, Tabita F., "Interactive control of Rhodobacter capsulatus redox-balancing systems during phototrophic metabolism", *Journal of Bacteriology*, 2001.

van Wezel A.L., "Growth of cell strains and primary cells on microcarriers in homogeneous cultures", *Nature*, 1967.

Wei, Qingyi; Lei Li, David Chen., "DNA Repair, Genetic Instability, and Cancer", *World Scientific*, 2007.

Index